Cat Confidential

www.booksattransworld.co.uk

CAT CONFIDENTIAL
VICKY HALLS

BANTAM PRESS

LONDON • TORONTO • SYDNEY • AUCKLAND • JOHANNESBURG

TRANSWORLD PUBLISHERS
61–63 Uxbridge Road, London W5 5SA
a division of The Random House Group Ltd

RANDOM HOUSE AUSTRALIA (PTY) LTD
20 Alfred Street, Milsons Point, Sydney,
New South Wales 2061, Australia

RANDOM HOUSE NEW ZEALAND LTD
18 Poland Road, Glenfield, Auckland 10, New Zealand

RANDOM HOUSE SOUTH AFRICA (PTY) LTD
Endulini, 5a Jubilee Road, Parktown 2193, South Africa

Published 2004 by Bantam Press
a division of Transworld Publishers

A catalogue record for this book is available from the British Library.
ISBN 0593 05276 5

Typeset in 12/15.75pt Cochin by
Falcon Oast Graphic Art Ltd

Printed in Great Britain by
Mackays of Chatham plc, Chatham, Kent

1 3 5 7 9 10 8 6 4 2

Papers used by Transworld Publishers are natural, recyclable products made from wood grown
in sustainable forests. The manufacturing processes conform to the environmental
regulations of the country of origin.

Thank you for everything, Mum, this is for you

CONTENTS

ACKNOWLEDGEMENTS

Thanks to my agent, Mary Pachnos, for her invaluable help at the beginning to make this book a reality and to Helen Newey at Good Relations for introducing us. Thanks also to Francesca Liversidge and the team at Transworld for their support and enthusiasm. I would also like to thank those veterinary surgeons who have supported my behaviour referral practice over the years. It is safe to say that this book would never have been written if it hadn't been for all my wonderful clients and their cats. You all have a special place in my heart. A big thank you to Peter, the brilliant custodian of my own cats; I couldn't have done this without you. Thanks to Sharon Maidment, Diane Sexton, Pat and Richard Shoebridge, Ruth Yates, Lee Boulton, Jean Perry, Danielle and Frank Gunn-Moore, Valerie Walter and Mark Evans for giving me confidence and encouragement and to Sharon Cole for laughing in all the right places. I mustn't forget a real debt of gratitude to Janet Valentine, who started the whole cat-behaviour-ball rolling back in 1983. Lastly, a very special thank you to Nick Murphy, the best friend any woman could have, who never stopped believing I could write this book.

INTRODUCTION

CAT LOVERS, OR AILUROPHILES TO BE ABSOLUTELY CORRECT, like nothing better than to talk about their cats. They fill their homes with cat pictures and cat tea towels and cat books and cat calendars and cat jewellery. They live with the most pretty, handsome, clever and intelligent cat in the world and they have a thousand stories for anyone who cares to hear them. However, for most people these charming anecdotes are easier to narrate than to listen to. Not me.

I earn a living by visiting people in their homes, drinking tea from their cat mugs and listening to their stories about their cats. I never get bored. I love every one. Some are sad, some funny, but all are important to my work. I am one of a small band of missionaries who spend our days helping people understand their pets. I call myself a cat behaviour counsellor. Others call me the cat shrink, the pussy doctor (not sure about that one), cat therapist or behaviourist. Whatever the name, the job I do is the same.

Cat ownership has changed dramatically in my lifetime. When I was very young I had to 'borrow' my neighbour's cat and concerned myself with the pleasure of the friendship without having to dwell on the responsibilities of ownership. Cat food was limited to a number of canned products and Jenny, my part-time pet, used to have some tinned cat food and some scraps from the table. I have to say that the 'scraps' Jenny enjoyed, being a pedigree Siamese of impeccable breeding, used to be tinned salmon and double cream! The rest of the owner involvement, as far as I could tell from my child's perspective, comprised letting her in or out when she cried at the door and the occasional stroke or cuddle. A bit of chatting whilst making the dinner and the typical owner/cat relationship of the 1960s was established. There were far fewer cats around in those days and far fewer people. The family unit was strong, people worked hard without many of today's labour-saving devices, and watching television was a half-hour treat rather than a major pastime. Neighbours were friendly, people were trusting and life ticked along. I am sure that I am looking back with the eyes of a child but people and cats appeared to be living a relatively stress-free existence. Cats were being cats and people treated them like cats. Veterinary care was limited in comparison to today, with far less emphasis on preventative medicine and general care. People rarely needed counselling and cats didn't have psychiatrists; one could argue that they didn't need them.

So what went wrong? What happened in the space of forty years that changed everything? Consumerism? Social overcrowding? Stress caused by the high expectations of modern life? The breakdown of the family unit? I have asked myself this question a thousand times whilst sitting in a traffic jam on the M25 on my way to see another emotional wreck of a cat. I believe a combination of many factors has led to a totally different style of pet ownership and care in the twenty-first century. We work harder than ever and often live in social isolation. There are now more single people living alone than ever before, myself included. Yet we still have a need for company and an even greater need to love and be loved, so we acquire a cat. We then place that cat with several others in the household (after all we can't get enough of them and we have an awful lot of love to give) or put them in a territory full of other people's cats. Or, worst of all, we demand an emotional relationship that the cat cannot possibly hope to provide. Inevitably, any of these factors has an enormous impact on this small carnivorous species and it develops emotional or behavioural problems that have never before been seen or adequately explored.

That is why cat behaviour counsellors like me exist. We are merely specialists with patience and knowledge who help the owner understand the problem and deal with it effectively and sympathetically. When I first started identifying and tackling these issues there were very few practitioners of the art. Many were academic scholars of veterinary medicine, zoology or psychology and the founders of the science of behaviour counselling. I am fortunate enough to have acquired a vast amount of practical experience in this field at a time when few others were out there in people's homes resolving their problems. Since the early 1990s the profession has grown beyond belief. Hundreds of students study the science of pet behaviour at universities and through correspondence courses. There is a great interest in the subject through media exposure, predominantly television programmes such as *Barking Mad* and *Pet Rescue*. Despite all this, people still feel that I do a ridiculous job, with the two most frequent comments being 'How can you possibly train a cat?' and 'What could possibly go wrong?' Other comments I receive are probably not worth repeating but I have to accept that it is a difficult concept to grasp.

I have therefore decided to write this book to explain what I do and why people need my services at all. Throughout the chapters it will become apparent that I use a great many human analogies to illustrate various points. I adopt this style in my consulting to enable difficult principles of behaviour in a complex species to be understood from our own perspective. Parallel emotions can thus be experienced, and the owner then remarks, 'I've never really looked at it like that. How awful for him!'

I must however take great pains to point out that I don't condone anthropomorphism: basically, attributing human thoughts to animals. This is of course wrong (cats are not small people in fur coats) but most of us can't help doing it to a varying degree. It is often the basis of many inter-species relationships. If using examples based on it helps to resolve the pet's emotional issues then I confess I am guilty!

Some problems that owners experience with their cats are not medical. They cannot be treated with tablets or surgery; this in itself is a difficult concept for many people. What remains are the behavioural, psychological or emotional problems that owners will constantly excuse and compensate for but which eventually come to control the household so that disharmony reigns. I am frequently amazed at how tolerant we cat owners are.

Many times I have visited people who have been living with

a house-soiling problem for years. Suddenly, six years or so down the line, they decide they can bear it no longer and it becomes an emergency that has to be dealt with. Often they seek help after such a long period because they have only just discovered that help is out there. I visited a household in London once that had been experiencing a persistent problem with one of their Burmese cats. He had been spraying urine around the house for some time and they had come to the end of their tether. During the course of my consultation, I leant down and touched the wall just above the skirting board to point to the urine. As my finger touched the wall it disappeared into the plaster and ended up in the cavity beyond as the wall disintegrated. It was very difficult, as I recall, ignoring the obvious fact that I had just damaged my client's property. So now we know. It's official. Cat urine destroys plaster!

Another lady I remember with fondness requested my assistance with a toilet-training problem with one of her pets. As the meeting progressed and I discovered that this sweet little cat had been peeing in the hallway for sixteen years, I rather felt that this was probably the way the situation would remain. I have been accused of magic and witchery in the past but I'm not really up to miracles.

There is no doubt that problems identified and tackled quickly are easier to resolve. The fact that most cases I see concern behaviour that has been evident for years is testimony to the tolerance and forgiveness of the typical cat owner. We love them so much we can excuse almost anything!

Without exception the one thing that all my clients have in common is this enormous and generous love for their cats. It is a very specific kind of love. It is nurturing, self-sacrificial and utterly compliant. What the cat wants, the cat gets (with knobs on). Unfortunately occasionally something goes wrong and the cat behaves in such a way that it is quite apparent all is not right with his world. The owners are often burdened with an incredible sense of failure and guilt when they request my services. They feel that the problem has occurred as a result of some oversight or omission on their part and, somehow, they have let their beloved pet down. This is rarely the problem. If they are guilty of anything it is that they love their cat *too* much.

I had an initiation by fire in the early nineties when I was given an incredible insight into the 'behind closed doors' nature of the potential human/cat bond. The owner was a lady in her early forties; I shall call her Miss X because somehow that seems appropriate. She had requested my advice regarding her cat's dependency. Miss X was soon to be travelling away from home and there was a need for her cat to have a short stay in a cattery. Since the two had 'never slept apart in eleven years' she felt this might be potentially distressing for him. I remember I found this admission about their nocturnal habits rather disconcerting and prepared for the meeting at her home with some interest but, naively, no trepidation whatsoever.

She lived in a small flat on the second floor of a large Victorian building. When I entered the flat I was immediately aware that this was not a typical cat lover's home. No cat ornaments or pictures apart from one extremely grand portrait in oil of her beloved pet. As I went to sit down in the living room her cat was watching me closely and promptly nipped me hard as I put my hand into my briefcase to remove my notepad. It was afterwards that Miss X told me I was only the second visitor to enter her flat over the past year (the first being the gas engineer). As the consultation unfolded I was aware of the fact that it wasn't going to go my way. It is extremely important in my job to maintain control of the discussion, otherwise questions that need to be asked get forgotten and the history-taking becomes incomplete. However, on this occasion it was clear that Miss X virtually had a script in her head about her relationship with her cat and *nothing* was going to stop her telling her story to a captive audience. So I sat back and listened.

Miss X was a very intense and eloquent lady who related her tale under the close scrutiny of her cat. She purchased him as a kitten from a local breeder and when she brought him home she provided him with a small bed on the floor in her bedroom and allowed him to explore. The first night he was very fretful and she gently lifted him into her bed for comfort. It was apparent that this had remained the sleeping arrangement since. A great deal of what she was saying was typical of an over-attached owner. She did absolutely everything for him and would rather spend time with him than with anything or anyone. She always fed him first and gave him bottled water and freshly cooked fish every day. When she spoke of him she used very personal terms and attributed many human traits and feelings to him; it was sometimes hard to remember that she was talking about a cat and not a partner. It was when she started to talk about her incredible love for her cat that I began to feel a little apprehensive. She felt an enormous sense of responsibility for him since she had taken him away from his mother. She said she believed she was his god and that his world revolved around her and the world she had created for him. She told me with some anger that her own mother had indicated that she was obsessed with her cat. She denied this vehemently and said their love for each other was merely that of one creature for another. It was love in its purest form and the lack of physical or sexual desire made it very special. It was his right to demand food, love and a cuddle. It was also his right to sleep in her bed and she should never deny it. She was also tormented by the thought that she would be punished in the afterlife if she died before her cat. This created a further dilemma for her, as she knew she would never be ready to lose him. Apart from a terminally ill mother in a local nursing home she had no other focus in her life.

This will always remain the most upsetting case I have ever had to deal with. Over the years I have seen a lot of cats in extreme distress and witnessed many owners crying over their problems with their pets. But Miss X's life has taught me a very important lesson about the concept of the human/animal bond. We can all experience the most incredible sadness, despair and sense of isolation. A pet cat can be an important but potentially fragile crutch. They are largely ill equipped to cope with this.

I tried to provide Miss X with sensible and positive advice about encouraging a sense of self-reliance in her cat by promoting other activities for him outside the relationship. I had absolutely no hope whatsoever that this would resolve their complex problem. I hadn't heard from her for several weeks when she made a telephone call to the veterinary practice where I was then working. She told me she was calling from her mother's room in the nursing home. In the background I could hear a man talking and Miss X told me that it was the priest giving her mother the last rites as she approached death. She said I was her only friend and she wanted me to be with her when her mother died. I will never forget that day.

Not all of my stories are quite so tragic. Mostly they focus on the positive things that can be done in even the most difficult situations. My office is festooned with cards and gifts from owners who, almost without exception, make the comment that I have changed their lives. Unless you have experienced the distress of a loved pet 'gone wrong' it is difficult to comprehend the enormous strain it can put on relationships and day-to-day living. One couple, Joy and Ian, were struggling with a house-soiling problem that had dominated their lives for two years. They loved their little cat, Whiskey, dearly but he had started to behave badly. He had taken to the delights of urinating behind the sofa and no amount of punishment, reassurance or tin foil deterrents made any difference. Their social life had plummeted (who wants to have drinks and witty conversation in a room that smells like a urinal?) and their own relationship had suffered from frayed nerves and a sense of divided loyalties. Ian, like any man, wanted a solution that was black and white: 'If he can't be fixed, rehome him or put him to sleep.' Joy could not even contemplate the implications of either option and in reality neither could Ian. They were both in utter turmoil and living in frustrated isolation from all their friends who wondered why they had withdrawn from circulation.

One day Joy read a magazine article about my work and immediately called me. We had a long chat on the telephone and suddenly Joy's spirits lifted as she saw the light at the end of the tunnel. I made no promises other than that I would do my very best. We met later that week after she had consulted her veterinary surgeon and arranged for a urine sample to be analysed to check there was nothing medically wrong. The vet reported that Whiskey was physically well but after spending some time with him I realized he was, emotionally, very worried indeed. He had an ongoing battle with a neighbour-hood cat and it would often stare at him through the patio doors and bang on his magnetic cat flap to try to get into the house. Whiskey was terrified of him and his nerves had gone straight to his bladder. He was retaining urine to bursting point and, when he was seeking sanctuary behind the sofa one day, his bladder just had to let go. This was the start of the whole problem and he had returned to the same place ever since whenever he needed to urinate because he felt safe there.

I devised a plan. With the aid of strategically placed litter trays, the removal of the soggy section of carpet and a little play therapy with a feather on a stick we were well on the way to a happier Whiskey. Joy and Ian were amazed; of course they had tried litter trays before but they just hadn't put them in the right place. Over the next few weeks, following all the advice I gave them, Whiskey became relaxed and playful and they fell in love with him all over again. They felt their lives were transformed as they started to see their friends and restore a degree of normality to the household. Every Christmas Whiskey and many other 'ex-patients' send cards; I I am probably one of the few people on this planet who receive more season's greetings from cats than humans! It's one of the many rewards for doing this incredible job.

Thousands of cats and their owners have touched my life over the years. Without exception they have all taught me individual lessons and every day, still, I learn something new. I hope this book will encourage you to really appreciate the nature of your relationship with your pets and to cherish them for it!

CHAPTER 1

The New Kitten

Annie's Story

ANNIE WAS NOT THE FIRST CAT I EVER OWNED BUT SHE WAS the very first kitten. As with most things in my life I adopted an 'initiation by fire' approach to learning and decided that a well-adjusted domestic kitten from a loving home was just too easy. Surely it would be more challenging and beneficial to play surrogate mother as well? I was working with the RSPCA at the time when Annie came into my life. It was a shelter funded by the goodwill of the local people and it was always busy. I remember distinctly that a gentleman came in one morning clutching an old sweater containing a creature of some sort. We were quite used to everything from seagulls to snakes arriving in this way and usually took bets on the species the garment contained. In this particular instance it was a thousand fleas with a small ten-day-old black and white kitten attached to them. The poor mite was in danger of dying from anaemia from all the blood-sucking parasites to which it was playing host so my first job was to remove as many as possible and provide warmth and nourishment in the form of powdered kitten milk. Little Orphan Annie (Annie for short) had been found all alone in a field and the kind gentleman had driven some distance to bring her to us. My boss at the shelter, Rex, felt it would be good practice for me to hand-rear her. Looking back I can see how true it is that ignorance is bliss, but I really didn't fully appreciate the enormous responsibility of providing a tiny new cat with a good blueprint for life. After all, I was far too busy feeding her at one end and inducing the necessary bodily functions at the other with the aid of warm water, cotton wool and a degree of gentle persuasion. At the time we were sharing our home with an elderly male cat called Hoppy (you'll meet him later) who had agreed to take over the running of the household for the princely sum of a seat by the fire and a limitless supply of prawns. Hoppy became invaluable in the rearing of Annie as we discovered his extraordinary paternal instincts towards the tiny newcomer. He soon took over the toilet arrangements and bottom washing (that was definitely the last time I let him lick my face) and became a feline swing and activity centre for Annie as she grew. I can see him now, trying to look dignified with a small black and white kitten dangling from one ear like a piece of jewellery.

Hand-rearing kittens

Hand-rearing is never the ideal start for any kitten but some-times it is essential for the survival of orphans or those whose mothers reject them. A better alternative would be to find a surrogate mother with a litter of a similar age since these females often accept other kittens and rear them as their own. In the absence of a feline family the responsibility rests on the shoulders of humans to do a half-decent job. Rearing single kittens can often be a problem as they are automatically deprived of the ability to interact with siblings or other kittens of a similar age. Young cats also learn a great deal from their mothers. Cats are extremely effective observational learners, able to learn new behaviours by observing another cat's actions. I wonder what they learn by watching us? Single kittens can lack social skills in later life and find it difficult to interact with their own species. Annie often seemed to get it slightly wrong when she bumbled up to my other cats and got a cuff round the head for her trouble.

There is also a rather disturbing incidence of adult aggression in hand-reared cats. Research indicates that this could be down to the inability of humans to accurately mimic the behaviour of the feline mother during the weaning process. As a kitten starts to eat solids there are going to be occasions when he still returns to mum's milk bar for top-up feeds. These are allowed or rejected at the whim of the mother and this teaches the youngsters a very important lesson in life. Frustration! You don't always get what you want and you have to learn to deal with it without going into a complete temper tantrum. If the process hasn't quite gone according to plan when a human and a syringe play the role of mother then things can become problematical. Every time the adult cat doesn't get what it wants it cannot look up the rule book on 'how to deal with frustration' because it never learnt it as a kitten. The results can be painful and disappointing.

Apart from the practical demands of being a surrogate mother there are many other considerations. When you see the incredible amount of growing that a kitten packs into a few short weeks, both physically and behaviourally, it is hard to imagine a more demanding and significant role than 'mother'.

Early development

Just take some time to remind yourself how quickly a kitten develops.

From birth to two weeks

Kittens' responses are limited during the first two weeks of their life and their very existence is totally dependent on the mother. They respond to her warmth and her touch and there is a strong instinct to find a teat using scent. Kittens often form a preference for one particular teat, their acute sense of smell enabling them to do so. At this age the kittens are relatively immobile and they can only use a slow paddling movement to travel very short distances around their mother and within the nest area. For up to three weeks the kittens are totally dependent on the mother's milk for nutrition. All nursing is initiated entirely by the mother. During the first two weeks of life the eyes will usually open at some time between seven and ten days, although any time between two and sixteen days is normal. Teeth are also starting to erupt at about two weeks of age to prepare the youngsters for rather more challenging mealtimes.

From three to four weeks

During the third and fourth week the tiny kittens' vision starts to play a role in guiding them towards their mother, rather than relying on her warmth and smell alone. A rather staggering walk appears during the third week and by four weeks of age the kittens can move a reasonable distance away from the nest. Around this time they start to develop the body-orientating reaction that will enable them to right themselves in mid-air when they are falling (such a useful technique for all adventurous felines). Under free-living conditions, mothers start to bring live prey to their kittens from four weeks after birth onwards to enable the youngsters to experiment with the manipulation and consumption of prey. Four weeks is also the age at which kittens normally start to eat solid food (or at least walk through it and inhale it up their noses).

From five to six weeks

By the fifth week the kittens are all over the place, showing brief episodes of running, and by six weeks they have started to move like mini-adults. As the weaning progresses the kittens become increasingly responsible for initiating bouts of nursing, not all of them met with compliance from the mother. By this time voluntary elimination has developed, and kittens are no longer dependent on their mother to lick their perineum to stimulate urination. (This is the time when it's essential to get the litter training right.)

From seven to eight weeks

Kittens have begun to show adult-like responses to threatening social stimuli. They will run, freeze or show aggressive behaviour just like mother to scary sights, sounds and smells. Weaning is largely completed by seven weeks after birth.

From nine weeks onwards

Complex motor abilities, such as walking along and turning around on a narrow fence, still take time to develop and may not be fully effective until ten to eleven weeks after birth. Visual acuity continues to improve until twelve to sixteen weeks. Sexual maturity can occur from six months of age (occasionally even earlier) and social maturity (adulthood) at any time between eighteen months and four years of age. There you have it! The kitten is all grown up in the space of two short years.

Social play

Social play with siblings plays a role in the development of later social skills. Kittens with no experience of siblings when young do eventually form social attachments but are generally slower to learn social skills than normally reared kittens. Solitary kittens also do not learn to hold back their bites in agonistic play behaviour if they target human hands rather than siblings. A person cannot possibly teach the boundaries of acceptable levels of physical force as well as another kitten.

Social play becomes prevalent by four weeks of age and continues at a high level until twelve to fourteen weeks, when it begins to decline. Social play fighting can sometimes escalate into serious incidents, especially during the third month. Play with objects develops at around the time when live prey is being introduced to the nest by the mother and kittens start to gain the eye–paw co-ordination that enables them to deal effectively with small, moving objects at around seven to eight weeks after birth.

Social play mimics agonistic social behaviour and predatory behaviour. There is however no evidence to prove that play increases successful predatory behaviour in adulthood. The latter appears to be influenced by observation of the mother, experience with prey when young and possibly competition between litter mates in the presence of prey. Despite early influences most cats become competent predators, albeit with particular preferences for the type of prey.

The development of a cat's personality

Anyone involved with cats will know how each individual has a unique character. The beauty of my job is that no two cats or problems are the same. Adult cats and kittens show considerable variation in their friendliness towards humans, whether familiar or unfamiliar. Even kittens from the same litter can differ markedly in this respect. If a group of six kittens, for example, is observed it will soon be apparent that one is shy, a couple are rather confident, one is exploring the room and knocking things over and the other two are seeking the company of humans and purring like trains. A different response to the same situation gives each cat its own unique personality.

The importance of early socialization
A cat's personality is developed as a result of both genetic and environmental influences. Genes 'programme' an individual with the potential to react in a certain way in certain circumstances. The individual's life experiences then influence whether that behaviour is ever actually expressed and to what level. The most significant behavioural and emotional development takes place at a very early age. The sensitive period is considered to be between two and seven weeks of age. During this time positive exposure to humans and other species allows the kittens to potentially form social bonds with anything from

chickens to chihuahuas. This process is referred to as 'early socialization' and is very much the responsibility of the breeder. Kittens should ideally not be placed in their new homes until they are twelve weeks old. This enables them to spend as much time as possible with their mother and siblings, from whom they can potentially learn a great deal.

Exhaustive research has been conducted into this sensitive period and the amount and type of handling that is required to give the kittens the best possible start. The conclusion of the studies is that handling by a number of different people during this time will tend to increase the sociability towards humans. Positive exposure to environmental challenges such as noise, children, dogs, different locations and even car journeys will better equip the individual to cope with life in the future. Don't you wish, whilst you are driving to the vet and listening to the cacophony from the cat basket in combination with the heady aroma of a hurried bowel motion, that your cat had experienced car journeys as a 'toddler'?

Whilst the period between two and seven weeks is usually outside the owner's influence it is still important to stimulate the three-month-old kitten as much as possible with a variety of challenges since lessons and positive associations can be learnt at any age providing the appropriate genetic 'blueprint' is present.

Categorizing personality

There are numerous ways to scientifically categorize character and personality, but there are two basic models, 'excitable and reactive' and 'slow and quiet'. Variations in excitability and timidity may well be caused by inherited differences, such as the amount of adrenalin released when faced with a challenge. (Cats have an instinctive 'fight or flight' response to danger and this is fuelled by the release of adrenalin that pumps blood into the muscles and away from non-urgent places such as the gut.) This variation in response can be illustrated beautifully by trying a little experiment when you are next sitting in front of the television. Move your foot quickly across the carpet a couple of inches from one side to the other. If your cat is now airborne then he is 'excitable and reactive'. If he is still lying on his back snoring he is 'slow and quiet'. Get the picture?

Certain breeds are often described by their temperament. For example, Siamese are considered to be sociable, affectionate, sensitive and vocal. They are also prone to eat wool, spray urine on their owners to get their attention, and pull their fur out, but that's another story. Burmese are traditionally assertive and outgoing with a tendency to be aggressive and territorial, and the Persian is placid whilst having a very tenuous grasp on toilet training. Such descriptions must imply that these characteristics are inherited. With regard to the behavioural traits it has to be said that these are personal observations based on my caseload over the last ten years and not derogatory remarks about three of this country's most popular (and highly delightful) breeds!

A new kitten – food for thought

I would never recommend that potential cat owners start with such a challenging task as a two-week-old but there are many months of fun to be had when sharing your home with an older kitten. It is possible to go to a pet shop, breeder or rescue centre and purchase a kitten with little planning or preparation. However, I don't recommend it! I would love to think that I could interest first-timers in the delights of cat ownership but it is a relationship that should be well thought out and planned for to reap all the potential benefits and avoid many of the pitfalls.

For example, we tend to take for granted that all cats will be clean indoors and that successful litter training is pretty inevitable. Whilst most cats remain exquisitely clean all their lives there are always those that fall by the wayside. Preferences can occur for rather inappropriate surfaces and if that surface is your goose feather duvet then I am sure you would do everything in your power to understand your cat and get the toilet habits right from day one.

Daisy and Puff – toilet training

The case I am about to describe was a joy for several reasons. It illustrates how badly things can go wrong and yet, with the right advice, resolve relatively quickly. It also was one of the few cases I have ever seen involving that delightful breed the Abyssinian (they just don't misbehave in my experience). From a personal point of view the house was gorgeous. How I love to wander round the homes of the rich and famous! It was a beautiful and graceful place in central London built over six floors. The basement consisted of several family rooms and the ground floor was the kitchen and informal dining room. The other four floors were formal entertaining rooms (wow, even I wasn't allowed in some of those), bedrooms and bathrooms. I remember getting a really good workout that day going up and down the stairs surveying the damage on the various floors caused by that most invasive liquid – cat urine.

Daisy and Puff were two delightful and friendly female

Abyssinians. They were sisters of about five months old and they had been purchased at twelve weeks by the family as pets for their two small children. Needless to say the kids were more excited by the carrier box the kittens came in (well, they were very young) and the care and entertainment of the new-comers fell on the shoulders of Alice, their mum. She had appeared to do everything right. She had purchased numerous toys, scratching posts and food bowls and provided Daisy and Puff with a covered litter tray in a discreet corner of the kitchen. The family had decided to keep the cats as indoor pets because they lived in a very busy part of town and they felt that there was plenty of room for two boisterous cats. They had agreed to keep the formal rooms 'out of bounds' but the stairs alone represented a huge area for fun and frolics. The kittens were very bold and inquisitive and they were soon exploring all over the house while the family laughed at the thunderous sounds that eight tiny feet could make up and down the stairs.

After a few days the laughter waned somewhat when a small pool of urine was found on a bath mat in one of the bathrooms on the third floor. Mutterings of 'babies' and 'accidents' followed and the incident was forgotten until, a couple of weeks later, one of the kittens peed on the duvet on the master bed during an excitable early morning game. Alice and her husband were not amused, and were even less so when several pools and then piles were deposited on other beds, sofas and carpets during the course of the next few weeks. In an attempt to arrest the development of this problem, the kittens were soon confined to the ground floor and basement and provided with a further litter tray in the kitchen. Unfortunately Daisy and Puff continued to use the carpet in the basement as a random toilet arrangement and Alice called me in desperation.

It was easy for me to see the problem and the story provided an important lesson about those crucial first few weeks in a kitten's new home. To a certain extent the desire to eliminate in a loose substrate is pre-programmed in every little domestic feline brain (their origins lie with the desert-dwelling African wild cat). When kittens start to control their own bowel and bladder function it is a relatively easy process for the mother to teach them the transition to voluntary elimination in a litter tray. If you watch a tiny kitten in a litter tray it is remarkable how it seems to know what is expected of it. This is the first step in forming a habit or conditioned behaviour to use litter, or similar loose rakeable substrate, as a toilet. Eventually there will be little conscious effort to decide where to go; it will be as easy as our trip to the bathroom in the dead of the night whilst still half asleep.

If you visualize the formation of pathways in the brain to create this sort of habit in an analogous way it may help to explain Daisy and Puff's confusion. Habits are pathways that are well trodden; they are footpaths in the brain's undergrowth that are so frequently used they are wide passages of bare earth. There is no question which way to go. When you create a new footpath there is a great deal of vegetation to cut through and the way forward is less well defined. When you are travelling down a new footpath it is easy to veer off in another direction and form new routes. Daisy and Puff had a new foot-path in their brain that led to an appropriate toilet but there were occasions when, three floors up and having fun, they weren't entirely clear which way it went when their bladders were full. So they had to make it up as they went along and, when the surface changed underfoot to a yielding one, their brains told them it was time to pass urine or faeces. Every time they felt the same change in surface they would think 'Toilet!' Hence a new pathway was formed and the more it was used the more likely it became that they would go down it. Can you see how young kittens can easily develop other surface preferences at an early age?

I explained this principle as best I could to Alice and she suddenly realized that the freedom she had given them had, unfortunately, not been the best plan after all. I devised a programme that concentrated on re-educating the brains of these tiny creatures to eliminate *only* in the presence of a suitable litter material. Sadly I can't talk things through with my patients so I have to adjust their surroundings to create situations that cause them to display the desired behaviour. In this case we needed to make a litter tray the most obvious choice when they needed to pee. Alice continued to confine Daisy and Puff to the ground floor and basement (an area that was larger than most family homes anyway). She provided a total of three litter trays, one in the kitchen and two in separate areas of the basement, using my favourite calculation of 'one tray per cat plus one'. I removed the flap entrance from the covered trays – we didn't want to make it too difficult – and experimented with a large open tray in the corner of the basement. Alice provided a lovely fine sand-like litter material (after all they are desert-dwellers at heart) that they seemed to adore immediately.

Providing enough of the right sort of trays in attractive locations, with the most desirable litter, was still not sufficient to deter the youngsters from travelling down their new brain pathways to the basement carpet. Residual smell of previous eliminations is also a strong trigger to perform again so it was essential that we removed any odour from the little patches in the corners of the rooms where they had previously soiled. This in itself is a bit of a task since there are so many 'odour removers' on the market it is easy to get confused and end up with a sodden mass that no longer smells of urine but of something far more obnoxious. Many heavily soiled sections of carpet need destroying together with the underlay. However, if the problem hasn't been too long-term the area can be cleaned relatively effectively (see Chapter 5 for a more in-depth look at inappropriate urination).

So we cleaned previously visited areas and we changed things round in the basement to make it look new and exciting. These kittens were learning loads of stuff about life and their surroundings and I wanted to make sure they had a stimulating and entertaining environment. The plan was to keep the kittens confined for a period of eight weeks and ensure that every single 'pee' and 'poo' went exactly where it should. (I shall never forget having a surreal conversation with a great ailurophile, Princess Michael of Kent, about cats' bodily functions. I was pontificating about 'urine-this' and 'faeces-that' when she suddenly said, 'Oh, come on! Pee and poo, dear, pee and poo.')

Anyway, time went on and Daisy and Puff were performing beautifully. It's amazing how consistent visits to the litter tray can provoke a round of applause; you just won't understand it if you've never had this problem! Gradually we allowed access to other parts of the house, with the exception of the formal rooms of course, and their preferred routes were starting to look like a favoured right of way for ramblers. I didn't push my luck and suggested that Alice place an additional tray in a discreet corner on the top floor, just in case. Bingo! The tray was christened on the first day and there were no further misdemeanours on duvets or bath mats. Daisy and Puff are all grown up now and still perfect angels.

A new kitten – making the decision

If the complexity of these little creatures hasn't overwhelmed you already you may be thinking of acquiring your very first kitten. Every new owner should ask themselves the following questions before embarking on a relationship that could span two decades.

Can I afford a cat? Whilst many people believe that it is merely enough to feed them there are a number of other significant costs involved. For example, vaccinations every year, regular worming and flea control, neutering surgery, microchipping and veterinary care for illness or accident. Not to mention the cat toys, 'cataerobic' scratching posts, fluffy leopard-print cat beds, radiator hammocks, fish tank, diamond-encrusted collars . . . you see how easy it is to get carried away?

What provisions will I make for family holidays? Catteries or house-sitters cost money and holidays need to be planned for many months in advance if you intend to use professional services. If you get very attached to your cat and figure he cannot possibly cope without you then you may have to abandon plans of going on holiday at all!

Do I live in a cat-friendly environment? Not all of us live in rural surroundings with idyllic hunting grounds away from the dangers of traffic. In reality most of us live in built-up areas adjacent to a road. Many live in flats or apartments that are not on the ground floor. Is it safe to let your cat out at night? Is it unkind to keep your cat confined to a small flat all day? This is a huge subject but you may find your question is answered later in Chapter 4.

What do I know about the cat as a species? Do people ever ask themselves this question before they acquire a cat? I doubt it, but it does make perfect sense to find out a bit about the animal you will be sharing your home with.

A trip to the vet

Regular veterinary care is essential for the modern cat but many prospective owners are deterred by tales of woe from other cat lovers about Sooty's complete phobia regarding any-one with a white coat and anal thermometer (I'm sorry but cats will not hold a thermometer under their tongue when you take their temperature).

It really is amazing that there is one word every cat seems to understand and that's *vet*. You just have to mention it, even in a whisper, and it heralds the immediate escape of the intended victim through the cat flap, not to be seen again for the rest of the day. Is it really 'that word' they are tuning in to or some other signal of an impending journey? The visual stimulus that creates this flight response is the cat basket, every time!

The biggest problem we are contending with is that most of the time the cat carrier is stored away in a shed or cupboard and brought out only when the cat is about to have a bad day. This usually involves a visit to the vet and the resulting trauma of a physical examination or a painful experience. Equally dis-turbing for some cats, it may mean a stay in the cattery and a temporary removal from everything that is familiar and secure. Either way it certainly means a trip in the car, which causes many different sensations in the average cat, all of them pretty negative!

The best way to tackle this problem is to plan ahead. A great deal of suffering in the future can be prevented if a new kitten is given as wide an experience of life as possible. The cat carrier can be left out at all times and lined with a furry and enticing bed. Instead of a portent of doom, the basket then becomes a friendly little resting place to curl up in after a hard day chasing butterflies. Regular short journeys can take place in the car, arriving back at home without any unpleasant experiences. The youngster then grows into an adult cat that tolerates car journeys and is less likely to view the basket as a dangerous thing.

This is great in hindsight, but (for existing cat owners) what if your cat becomes a tiger when the basket appears and no amount of furry bedding will entice him? Such a cat has learnt a strong association between the cat carrier and danger and his highly efficient survival instincts tell him to get as far away as possible when it suddenly appears. Once again the key is to plan ahead. If the visit to the vet is for a routine check-up or vaccination then the basket should make its appearance a week before. Very casually it should be placed in a well-frequented room and left with the front open. This will cause an immediate adrenalin rush in the cat and he or she will probably disappear. Mealtime will come and every-thing in the household will be confusingly normal. No hushed voices, no urgent or anxious body language, no humans making a grab for the scruff of the neck . . . just the basket. Days will then go by and a cat can only stay aroused for so long. When animals are exposed to a previously negative stimulus and nothing bad happens, they gradually become less and less reactive to it until it no longer means anything. That is, until the next time!

There is another possible technique that takes advantage of the cat's powerful sense of smell. Scent is an extremely im-portant means of communication for our feline friends and certain subtle odours can affect their mood. When your cat makes a loving approach towards you and rubs his face on your hand he is depositing one of these messages. There are glands in a cat's cheeks that produce a secretion called a pheromone: a scent that carries specific information for other members of the same species. The cat's facial pheromone transmits a positive and secure feeling and a happy and confident cat will 'cheek rub' furniture and doorways throughout the house. There is a product available in spray form from your veterinary practice that is a synthetic version of this naturally occurring pheromone. If it is sprayed lightly inside and outside the basket about half an hour before your cat is placed in it, there will be an overwhelming sense of familiarity and security when he steps inside. This may just make the trip to the vet's a little less stressful for both of you!

Planning family holidays

Visits to a good local cattery during family holidays don't have to be too traumatic for your cat providing he or she has been introduced to the concept of regular visits from day one.

It is essential to plan ahead since all popular catteries get booked up many months in advance. When choosing a cattery it is advisable to view first rather than just make a decision based on the fact that it's called 'The Posh Pussy Hotel For The More Discerning Client'. Visit the cattery during their published opening times or phone first and make a specific appointment. Ask to be shown the facilities and pay particular attention to the security of the premises and the overall cleanliness. A smell of disinfectant could be a warning sign: it may be an attempt to mask a general lack of good hygiene. The cat enclosures should be large enough to provide an enclosed sleeping area and an open-air exercise run with barriers between each pen to prevent the airborne spread of infectious diseases. The availability of heating and cooling equipment should be checked to ensure that the temperature within the enclosed sleeping areas can be controlled for the comfort of the resident. The cattery should be well maintained and the concrete areas free from any green algae stains. The food and water bowls provided for the cats should be clean and the litter facilities regularly checked for soiling. When you are viewing the premises it is also advisable to check out the residents. Do they look relatively content?

A good cattery owner will ask questions about your cat to ensure he or she is regularly vaccinated. They will also be happy to discuss special dietary arrangements and any other specific needs that require extra attention. Only when you are satisfied that all your requirements can be fulfilled should you make the booking. The Feline Advisory Bureau provides guidelines to cattery owners with regard to standards of care and it is a good sign if the chosen business carries their approval and is mentioned in their listing.

Choosing the right cat for you

Fundamental issues have been raised and plans have been made. Now the acquisition of the new family member can begin. There are, however, always more questions to be asked to further confuse your already befuddled mind. At this stage you need to decide whether you want a kitten or an adult cat. Should you give a home to a moggy or a posh pedigree? You even need to think about what type of fur your future companion should have.

Kitten versus adult cat
Kittens tend to be a better choice for families that already have another cat in residence. (Please read Chapter 5, however,

before deciding whether Tigger is actually lonely or thoroughly enjoying his single status.) Kittens are great fun but they can be a handful when they are little and many breeders and rescue centres stipulate that two kittens should be rehomed together, especially if you are out at work during the day.

Second-hand adult moggies have established personalities that they may actually allow you to glimpse in the confines of a pen in a rescue shelter. If you are very lucky the purring cutey will get home, behave exactly as expected and become the perfect pet. If you are unlucky he will turn into a spitting monster that remains under the spare bed for a week and then uses the house as a hotel once he has found the cat flap. Let's face it, this is a lottery. All my cats, with the exception of Annie and Bink, were pre-owned, some of them for several years, so I can safely say I have played that lottery and won every time. There are countless adult cats of varying ages (isn't it funny how all cats in rescue catteries are three years old?) that desperately need a good home, but don't be shy about finding out as much as possible about their previous lifestyle.

Pedigree versus moggy

To a certain extent, most breeds have typical characteristics which allow prospective owners to choose a highly sociable Burmese, a mischievous Devon Rex or a placid Persian and all the variations in between. It is still important to ensure that the kittens are appropriately socialized with people and other animals and that the breeders have provided them with a good basic 'education' in the joys and perils of domestic life. Pedigrees, on the down side, have a tendency towards inheritable and congenital diseases and all potential risk factors should be discussed with your veterinary surgeon or practice nurse before making the final decision. Domestic short or longhairs (fancy name for moggies), on the other hand, are pretty much as nature intended and they can be as loving and beautiful as any expensive pedigree.

Long-haired versus short-haired

Short-haired cats are less prone to the formation of fur balls (from grooming and ingesting fur) and are perfectly able to maintain their coat condition with their tongue.

Long-haired cats, particularly the pedigrees, need daily grooming and many (ironically) hate the entire process. It is often impossible for some owners to keep these coats from matting and a visit to the vet's practice becomes necessary. The cat then suffers the regular indignity of wandering around looking like a poodle having a bad hair day until the fur grows back. Semi-longhaired cats tend to cope much better with their coat and can keep it fairly free from mats and tangles.

Heart versus head

Choosing a kitten should always be a labour of the mind rather than the heart. We tend, however, to view a litter of kittens and make a decision based on 'he looks just like a cat I had twenty years ago' or 'that one isn't playing with the others, I feel sorry for her' or 'I've always wanted a grey kitten'. Whilst these methods are as useful as tossing a coin in the art of decision-making, it is still entirely possible that you will have a long and enjoyable relationship with your cat. However, to increase the likelihood of success it might be better to bear the following advice in mind.

Research the litter by asking questions before you view. An unsuitable litter is easier to reject before you see their little faces. If the answers to any of the following questions raise doubts in your mind then it may be better to look for another breeder:

- Have the kittens been reared in a domestic home or a pen outside?
- Have the kittens been handled by a number of people from the age of two weeks?
- Can the mother be viewed with the kittens?
- What is the mother's temperament?
- Have the kittens been examined by a vet and have they been wormed and treated for fleas?

If the kittens are in a rescue centre, then it is important to find out as much as possible about their background. Do not be tempted by feral kittens that spit and hiss and are reluctant to be handled. They will be very hard work and should only be taken on if the family as a whole understands they may never be the ideal friendly pet.

Decide beforehand whether you are looking for one or two kittens. Single kittens are advisable as companions for existing cats in the household. If you work during the day then a pair of kittens will be good company for each other when they are growing up. Choose litter mates that appear to be close and playing together.

When viewing the litter look for a kitten that fulfils the following criteria:

- Bright eyes with no discharge
- Clean anus with no sign of diarrhoea
- Clean ears with no evidence of dark brown wax
- Shiny coat and no pot belly (this would indicate a worm burden)

- Alert and interactive with the environment
- Playful with the other kittens in the litter
- Keen to approach visitors

Planning ahead

Once the decision has been made the newcomer can be brought home. However, stopping on the way to collect a litter tray, food bowl and sparkly collar is hardly consistent with the need to plan ahead. Before your new kitten comes home a suitable room should be equipped with the following:

- Soft bed (try to bring home some familiar bedding with the new kitten for comfort)
- Food bowl and separate water bowl
- Litter tray and litter (maintain the same type that the kitten has used previously to avoid confusion)
- Cardboard boxes (for hiding and playtime)
- Toys (some toys can be left out but anything dangling on a piece of string should only be used when you are around)
- Scratching post

If you are adding to your existing feline family then please take a moment to consider this next case before you plan the introduction.

Boris and Chivers – introducing a kitten to an established cat

Kathy and Trevor and their two young children, Tamsin and Harry, lived in south London. They had, until recently, shared their home with two eight-year-old moggies called Boris and

Karloff. They were both sleek black short-haired cats and they made a formidable team as they patrolled their territory. Despite their appearance they were soppy daft things with the family and apparently devoted to each other. Tragedy had struck three months previously and Karloff had fallen victim to a speeding car in the normally quiet side road where they lived. The family were devastated, particularly Tamsin who was Karloff's favourite companion. The poor child was inconsolable and her sorrow was only matched by that of Boris, who howled and paced around the house searching for his lost brother. Several weeks went by and Tamsin and Boris stopped crying but continued to mooch about the house with sullen faces. The family had a meeting and decided that Boris was lonely and needed a new companion.

Kathy and Trevor were sensible people and they planned the new addition with great care. They visited the local rescue centre and viewed a number of likely candidates for the position of companion for Boris. They eventually settled on an adorable eight-week-old male ginger kitten with a little pot belly. Kathy felt that it looked more like a furry satsuma than a potential replacement for a lost loved one but Tamsin was adamant so the papers were signed and the family took him home. They decided to call him Chivers and as they entered the house there was an air of tremendous excitement. They visualized Boris with a big smile and watery eyes as he rushed, open pawed, to greet his new friend. Well, it wasn't *quite* like that . . .

Kathy, Trevor and Harry went straight into the sitting room where Boris was sleeping, totally unaware of the new arrival. Boris woke, stretched, chirruped and moved lazily to plonk himself on Trevor's lap for an afternoon tickle. Kathy gave the agreed signal and Tamsin entered the room with the little ginger kitten cradled carefully in her arms. What happened next will be firmly etched in the family's memory for ever (sadly). Tamsin approached the relaxed Boris and gently lowered the chubby kitten until the two were face to face. In a split second Chivers emitted a fearsome spitting sound and doubled in size. Simultaneously, Boris became rigid, his eyes widened and turned very black and every single claw dug deep into Trevor's legs. Trevor and Boris screamed and then Boris threw a punishing left hook directly at the kitten. Chivers went flying out of Tamsin's arms and landed behind the sofa. Boris ran out of the room hissing, growling and generally swearing feline obscenities and the family froze. Kathy was the first to move and ran behind the sofa, terrified of what she might find. Chivers was lying on his side, breathing rapidly and shaking from head to foot. She picked him up carefully and put him in a blanket and rushed him to the local vet. Chivers was examined and kept overnight for observation. The family returned home; none of them was able to look Boris in the eye as he wandered into the kitchen looking as if nothing had happened and expecting his dinner.

Chivers recovered fully and the vet decided that he should come home. The family had another meeting. They all agreed that they couldn't possibly take Chivers back to the rescue centre. They desperately wanted to make it work but they felt they needed expert help. Luckily Kathy's friend had sought my advice some months before and she had said that the process had been extremely successful. It was decided that I would visit and help the family organize a more diplomatic introduction.

I asked them to hire a large 'kittening pen' from the veterinary practice to help with the programme of introduction. When they brought Chivers home he was confined initially in a single room where he had food, water, toys, a litter tray and a bed. These were all provided within the cage, with the door left open to allow him to explore the rest of the room. I explained that the choice of room was very important if Boris was to be kept on our side. It had to be a place where he didn't spend any time so that he wouldn't feel instantly excluded from a favourite area. The family chose the music room at the back of their rambling Victorian house. The children's musical talents were not quite as expected and they spent little time practising on the piano. As the family didn't go in there often neither did Boris so it was agreed that this would be the ideal location for Chivers to settle into his new home.

Chivers adjusted well to his new environment with no apparent after-effects from his recent trauma. Tamsin spent a great deal of time playing with her new friend and he enjoyed the cage as a cosy place to eat and curl up for much needed rest. Whilst little Chivers was being entertained, Boris was equally pampered in the rest of the house. It was important for the family to gauge the amount of attention that Boris required since it is a common mistake to lavish a huge quantity of 'touchy-feely' love on a cat that never wanted it in the first place. This certainly doesn't compensate for a new addition; it merely makes the whole situation more traumatic. Boris liked loving on his own terms but he did have a thing about ham, so it was decided that he would be adored on request and provided with tiny but delicious slivers of ham more frequently than normal.

This regime was followed for a whole week until the family was ready and willing to advance to stage two. Chivers was being fed five times a day and he thoroughly enjoyed his food. Boris ate dry biscuits that were available throughout the day in the kitchen but he did love a treat of a sachet of gourmet cat food or a little honey-roast ham. It was time for Boris to see

Chivers, in the safety of his cage, with no risk of injury to any member of the family if it all went pear-shaped like last time. During this period of introduction the family was asked to feed both parties simultaneously, with Chivers inside the cage and Boris some distance away in the doorway of the room. They duly enticed Boris with a small saucer of cod in prawn jelly and placed his treat in the entrance to the room in full view of the ginger kitten in his cage. Chivers was chomping his food noisily and paid no attention to the older cat whatsoever. The family collectively held its breath in anticipation of a slightly more demonstrative reaction from Boris. They were pleasantly surprised when, after fixing Chivers with a wide-eyed stare, he thought better of it and cleared his plate instead. Boris was over the first hurdle and well on his way round the track!

Life continued with a series of non-violent feeding en-counters, and each time Boris's bowl was moved slightly nearer Chivers's cage. After the first couple of weeks the family moved the pen to another room and started the process again. By the fifth week the two cats were eating within two feet of each other and Boris even started to wander into the room without the lure of food. He quite liked watching the kitten play. Kathy, Trevor and the children were patient and loyal to the programme and by the eighth week it was time for Chivers to enter hallowed ground – the kitchen. Once again Boris delighted them as he sauntered past Chivers's pen by the Aga and on out of the cat flap without even a glance. It was time to try a face to face. Harry and Tamsin were sent to their friends' houses for tea and Kathy and Trevor prepared them-selves for any eventuality, armed with a camcorder (to record a happy event) and a pillow (to break up the fight if the out-come was not so happy). The door to the pen was opened and the tension was palpable in the two humans in the room.

Oblivious to the significance of the occasion, Chivers, now a leggy four-month-old kitten, trotted out of his cage to explore the kitchen. He spotted Boris sitting in the corner by the cat flap and immediately walked over to him, reached up with his little ginger head, and plonked a tiny wet nose on Boris's face. Boris raised his paw and slowly squashed Chivers to the floor. Trevor had to restrain Kathy as she lurched forward with the pillow because he could see what was coming. Chivers rolled over and Boris flopped down and they started playing the most gentle and relaxed game. Boris and Chivers were officially friends!

Some years on, I have stayed in touch with Kathy and Trevor. Boris and Chivers remained good companions. Chivers grew up and became rather more independent but they still went hunting together and they still came home together. Boris is quite elderly now and Chivers will tease him occasionally but they can both still be found squashed into a sagging radiator hammock together in a ginger and black ball.

Introducing a new kitten – quick summary
- Kitten pen in rooms of least importance to resident cat first and then moved throughout the house.
- Allow the kitten to exercise within the room when the other cat is not around.
- Feed new kitten and established cat either side of the pen.
- Reduce the distance between the two cats when they are feeding by small amounts daily.
- Exchange bedding between the two to allow them to become familiar with the other's scent.
- Provide attention to the existing cat but do not exceed the amount that he or she finds enjoyable!
- Allow several weeks before opening the pen and letting the

cats get to know each other.

- Keep a cushion or pillow handy to place between them just in case things do not go according to plan (never separate cats with your own body parts).
- Be patient – a bit of extra effort at the beginning can make a lifetime of difference to a relationship.

❖   ❖   ❖

Annie is now a beautiful twelve-year-old cat. She absolutely adores me and has become more and more affectionate over the years. She mixes well now with the other cats (I couldn't resist a total of seven at one point) and has a habit of greeting me by leaping up on her hind legs to head-butt my hand. She is carrying a little extra weight these days and her legs are stiff and creaky first thing in the morning, but she will always be a kitten in my eyes.

CHAPTER 2

The Scaredy Cat

Spooky's Story

MY OWN PERSONAL EXPERIENCE OF LIFE WITH CATS BEGAN with Spooky, a young adult tabby and white cat with the most beautiful eyes I have ever seen. She was adopted from the local RSPCA shelter after many months of research and careful thought to ensure I gave a home to the right sort of cat for my lifestyle. I managed a thriving mail order fashion business at the time and I was rarely at home. I had wanted a cat ever since I was a small child but constant nagging and pleading with my parents had little effect. When I was about nine years old I smuggled my new neighbour's cat, Jenny, into my bedroom on many occasions just to prove to my mother that this was what

I really wanted out of life. She would lie under the bedclothes with me and I distinctly remember how impossible it was to stop her purring and giving the game away when my mother came into the room. Jenny was a beautiful sealpoint Siamese; the old-fashioned kind with a fuller face and robust physique. I loved her with a passion and truly believed that she loved me. When I got home from school I used to run straight out of the back door calling Jenny's name. She would trot down the garden to greet me with her 'bumpy miaow' that was interrupted rhythmically with each footstep. She used to leap into my arms (I was tall for a nine-year-old otherwise she would have bowled me over) and I would squeeze her and kiss her and tell her about my day. When it was wet outside and she had been hunting it was *my* door that she used to cry outside for me to towel her dry in front of the radiator. I used to love doing that; it made me feel important and parental. Who needed dolls when you had a Jenny! About eighteen months after Jenny came into my life, when I was ten years old, we moved house and I was devastated. My parents were excited about moving to a much larger house, conveniently situated for my school, but I wasn't convinced. I couldn't imagine life without Jenny but my mother and my neighbour insisted I could return and visit her regularly.

Shortly after we left Jenny went missing and never returned. She had been howling outside our back door constantly and then one day appeared to set out in search of her friend. That night my neighbour heard a fox and she believes to this day that Jenny was attacked and killed. I still feel I let her down somehow.

It is worth noting that there is a great deal of speculation and debate about foxes and the risk they represent to domestic cats. Many people report a variety of observations from genuine social interaction to actual hostility. There is some evidence to suggest that, every now and then, a female becomes skilled in the predation of pet cats. This may account for seasonal 'disappearances' of groups of cats in certain areas when the vixens are feeding their young. I cannot say for sure how great a problem foxes are in general but I don't take any chances. I can't help thinking of Jenny.

I remained in a cat-free household until I got married. I was now able to make a decision for myself about cat ownership but initially felt I wouldn't be able to provide a good enough home. I wanted to be the perfect owner and as I worked long hours I knew I couldn't give a good home to a kitten. The option of acquiring an older cat did not really dawn on me until a friend suggested it. So I read and read and tried to obtain as much information as possible about caring for cats. After seeing an advertisement in the paper I went for an 'interview' with the local branch of the RSPCA. I seemed to pass this interview with flying colours and was invited to visit a small potting shed at the bottom of a garden in rural Kent to try to find the right cat for me. The lady running the cattery, an enthusiastic and likeable lady called Daphne, steered me towards one particular cat that she was having difficulty in rehoming. A sweet little tabby and white female was sitting at the bottom of the shed with her back to everyone. She was breathing rapidly and was very tense; I could sense her distress. She had apparently been brought into the shelter as a stray with young kittens. The kittens had been rehomed and she had been spayed but had become increasingly depressed and withdrawn. As I picked her up I felt her stiff little body become even more rigid and I felt guilty that I was adding to her upset. I continued to talk to Daphne and, without thinking, started to gently rock the cat in my arms. Suddenly I realized that she had moved closer in to my body and that she had started to purr. Sold!

We called her Spooky (because she was) and took her home on a cold day in March. We provided her with a soft bed in the kitchen next to the radiator. My husband, Peter, went away on business shortly after she arrived so the job of getting her acclimatized fell on my shoulders. I took a week off work and simply lay on the floor next to her, talking to her. I stroked her and fed her by hand and still she remained fixed to her basket, only coming out in the dead of night to eat and use her litter tray. I felt frustrated that all my love was not being well received but during those first few weeks with Spooky I learnt a very valuable lesson. Just loving a cat is not always enough. We need to love them in the right way. In complete exhaustion, I ignored Spooky for a couple of days and went about my business as usual until one morning, whilst I was watching television and eating Marmite on toast, a little tabby and white cat sauntered into the living room and jumped up onto the sofa next to me. That day heralded the start of a wonderful love affair between Spooky and me and between Spooky and Marmite!

I didn't really understand at the time how Spooky came to be so nervous. I felt, as many people do, that she had been the victim of some cruelty. In this particular case, ironically, I was probably right. Spooky remained timid through a burglary, three house moves and the introduction of six more cats into the household. She loved routine and would thrive on just being and observing with the occasional cuddle. She was never aggressive to the other cats or to humans. If she was frightened she would freeze and become so rigid that you could turn her upside down and she would remain fixed like a wooden toy. She was the queen of sweaty paws. Living with Spooky has made me particularly sensitive to shy cats and I have seen and helped many over the years. It isn't enough in my business to label a cat as nervous or fearful without understanding the origins of the emotional state. It isn't necessarily the result of cruel treatment at the hands of humans as many people suspect.

Fear and anxiety manifest themselves in different ways. A cat can be anxious for example and it would be impossible to tell by just observing it at one particular moment. Anxieties are often internalized and the cat retains a relatively normal countenance throughout until something gives way and behavioural signs start to develop. Fear tends to be rather more obvious; I don't think anyone would be in any doubt about recognizing a truly fearful cat. Here are just a few examples of what you would expect to see:

Signs of fear
- Dilated pupils
- Rapid respiration
- Rapid heart rate
- Tense or rigid body
- Low crouched body posture
- Piloerection (raised fur on the back and tail)
- Sweaty paws
- Trembling
- Aggression
- Escape
- Hiding/freezing/avoiding
- Involuntary elimination

Signs of anxiety
- Tense body
- Lip licking

- Dilated pupils
- Hypersensitivity to noise/movement/touch
- Urine retention
- Urine spraying
- Inappropriate urination
- Over-grooming
- Change in normal routine/patterns of behaviour

I see many nervous, anxious and fearful cats in the course of my work. Some of these emotions can be resolved, others can only be managed as they form a fundamental part of that individual's character. The next case illustrates how difficult it can be for young kittens, deprived of early socialization, to overcome their fears.

Joey and Jessie – the nervous kittens

Many behavioural problems appear to develop at the onset of adulthood or the time when the cat is considered to be socially mature. This can occur at any age between eighteen months and four years. This is often the time, for example, when previously equable cats turn into solitary individuals who find the concept of living in close proximity with other adult cats positively disgusting or incredibly stressful. The reason why all mature cats do not react in this way is probably another example of variation in the degree of sociability with their own species.

Occasionally I am called to see young cats or kittens that have suffered from a lack of early socialization. There is a growing trend in urban areas to take feral cats off the streets and try to 'rehabilitate' them to domestic life with humans.

Feral cats are basically cats with a history of domestication, either their own or in previous generations, who are living a natural and wild existence. They are often dependent on a man-made food source but they are frightened and fiercely aggressive towards humans if cornered. Often the adults are impossible to rehome successfully and they are neutered and returned to their previous environment. The kittens, however, can adjust, but they often need extreme patience and care to enable them to live happily with humans.

One such case that was typical of many related to two kittens that were adopted by a caring single lady called Mary. She was working full-time and living in a ground-floor flat in London. She had always wanted cats and felt she had enough love and patience to cope with a couple of kittens from a difficult background. Joey (male) and Jessie (female) were approximately five months old when Mary first saw them in their foster home. They were born outside to a domestic mother who was well socialized with humans. Unfortunately she continued to stray whilst she was rearing her kittens, which meant that the latter had no apparent contact with humans during that most sensitive developmental period between two and seven weeks of age. They were extremely nervous and feisty youngsters when their mother was found so they were taken to an experienced foster carer to get them used to domestic life. Mary's first sight of them consisted of tips of tails as they disappeared behind furniture but she remained undaunted. Luckily she was an extremely sensible lady and she enlisted my help shortly after she brought the kittens home.

Despite having made progress in their first home, the kittens were extremely nervous on arrival at Mary's flat. She had been advised to keep them in one room initially so that they could acclimatize to small areas at a time. It is quite usual for new owners to choose the kitchen as the ideal location but there really couldn't be a more unsuitable place. A kitchen is full of cupboard units and electrical appliances all of which have tiny areas behind them that are totally inaccessible to humans. I can put my hand up and say that I too made this error of judgement when Spooky first arrived and spent an agonizing (for me) twenty-four hours behind the washing machine. Needless to say Joey and Jessie disappeared behind the fridge/freezer and put large holes in the fingers of a willing accomplice who helped Mary remove them. This was not a good start.

Poor Jessie and Joey had been deprived of the important opportunity as tiny kittens to face the challenges of contact with different things. They had developed into reactive young adults who found every new situation frightening and stressful. They just didn't have the previous experience to help them work out the appropriate response.

When I arrived the kittens had been shut into Mary's bedroom. I decided therefore to conduct the consultation on the floor of her bedroom in the hope that the kittens would emerge from their refuge under the bed and show themselves. (I have lain on many floors during my career, in homes ranging from palaces to council flats, since it is such a useful position from which to attract even the most reluctant cat.)

Although neither kitten had had early contact with humans it was soon apparent that both were naturally bold and inquisitive cats. As I spoke to Mary and focused intently on her, the conversation flowed and I had her undivided attention. This was a very non-threatening arrangement for the kittens because we were both completely ignoring them. Our presence became less dangerous as time went on and consequently far more interesting. When I removed from my briefcase a small fishing rod with a piece of string and a feather tied onto the end, their interest could no longer be suppressed. I started to agitate the feather beside the bed in a rather distracted manner and very quickly a burst of tabby erupted from its hiding place and 'killed' the toy. By the end of the consultation Mary and I were both sitting on the floor with our legs stretched out and the kittens were rushing up and down as we provided an exciting and challenging assault course. By carefully introducing them to the delights of human company and making every interaction pleasant and non-threatening, Mary ensured that Jessie and Joey gradually became less and less reactive to her presence. It was important that this process included other people so that the kittens did not develop a trust of one particular individual to the exclusion of all others. Mary enlisted the help of her friends – even the one with the punctured hand showed willing – to spend quality time with her cats. I taught her important rules relating to the interaction between humans and nervous kittens, including the need to ignore them and not focus on them by walking towards them or looking at them. Being ignored gave the kittens a sense of safety and made humans look less dangerous.

It is always important when dealing with fearful cats to avoid reinforcing that fear. Often these cats are responding to an everyday occurrence or item and the fear is quite unfounded. It is a natural instinct to comfort our pets when they appear distressed and say things like, 'Mummy's here. There, there. There's nothing to worry about.' Unfortunately our actions are being misinterpreted and if the cat had a supernatural grasp of English it would hear, 'Your protector is here. You are quite right to be fearful but I will protect you.' It is far better to go about your normal business, thereby signalling 'Crisis? What crisis?' – a much more positive message. It is a hard concept but it works. Mary was advised accordingly. She had previously pussy-footed around the kittens to avoid frightening them, but when she and her friends behaved perfectly normally they actually put across a much more relaxed body language. Think about what happens when you want to get your cat into a basket to go to the vet or give it a pill. Trust me, your walk and look change and that means trouble for your cat!

It was also necessary for Joey and Jessie to experience new challenges to help them build up their confidence in general, so we discussed a tailor-made therapy programme to get them started.

The kittens were allowed the run of the bedroom, hallway and lounge whilst Mary was in the house. We made sure that the open fireplace in the lounge was blocked and any other potential hazards had been removed from the area. The kitchen was out of bounds, given the previous unfortunate experience. When Mary was not in the house the kittens were left in the bedroom with the door shut. A tall modular scratching post called an activity centre was placed in the lounge in a location where the kittens, if they climbed it, could see all of the room and out of the window. The unit was extremely elaborate and included sections where the kittens could hide if they felt vulnerable. Cardboard boxes with holes cut in them were placed in the hall and lounge to give the kittens a safe place to hide away if they felt threatened. If either Joey or Jessie found a favoured hiding place it would be respected and not disturbed to ensure it continued to be perceived as secure.

Mary was asked to play with the kittens as often as possible, using toys on fishing rods so that the item they were playing with could start well away from her body. She promised to make no eye contact and encourage the kittens to play the games as near to her as possible. Once the kittens were distracted with a game Mary began to stroke each one, starting with the head and back and eventually progressing to even the most vulnerable areas around the belly. Dried catnip was also used to entice the kittens to explore closer to her. Catnip, or catmint, is my most useful tool in cat behaviour therapy. It has the most incredible effect on two-thirds of the cat population: they will sniff it, eat it, roll in it and give the most convincing display of ecstasy you are ever likely to see. Whatever else is going wrong in the cat's world it is forgotten when catnip is around.

Once the stroking was well tolerated then Mary could concentrate on developing the concept of restraint. It was no good having kittens that you could touch but couldn't pick up in an emergency. This process started with gentle holding pressure for a few seconds followed by immediate release and more stroking. Then it progressed to taking the weight of the kitten and lifting for longer and longer periods of time. Ultimately the goal was to be able to pick either kitten up and hold him or her against Mary's body.

Mary was an enthusiastic follower of the programme and after eight weeks both kittens had been introduced to the cat flap and were spending short periods of time outside when Mary was home. They were also approaching her regularly for attention and Joey had become a lap cat! Mary and I couldn't have been more pleased.

Cats and 'psychological warfare'

Anxiety in cats is not always exhibited quite so obviously as it was by Joey and Jessie in their early days. Cats have an incredible ability to internalize their emotions and stresses and withdraw into themselves if things get difficult. Just like us, cats have modern-day stresses and one of their major issues is social overcrowding. For those of us who have ever spent time commuting on the underground this must be an understandable feeling. If a cat is bothered by another outside in the garden or even one of its companions indoors it does not necessarily indulge in violent fighting. A great deal of aggression in the cat world is much more sinister. Fighting over every dispute does not make good sense for cats since they are armed with formidable weaponry, and whilst each individual can inflict life-threatening wounds on another it can also be injured itself. Adopting a strategy of posturing and psychological threat is far more sensible.

There is one method employed by many cats within multi-cat households to intimidate other members of its group that I find ruthless but enormously effective. Even under the closest scrutiny from a loving owner, little Sooty can wage psychological warfare against his adversary whilst looking cute and innocent. Have you ever had that well-known stress dream where you suddenly realize you are naked in a public place and everyone is staring at you? Or the one where you are a child again sitting important exams and you don't know any of the answers? If cats were to dream as we do, their stress dream would be this. Tigger is standing at the top of the stairs with a very full bladder. Sooty, his arch-enemy, his Nemesis, is sitting nonchalantly on the bottom step chewing his claws (and probably whistling). Tigger knows the litter tray is in the kitchen but he also knows that if he attempts to pass Sooty he may get bashed. Tigger also knows that Sooty knows that Tigger knows he may get bashed. Unfortunately the bladder isn't getting any emptier and Tigger is stressed and may well have to go where he stands.

Lola and Issie – anxiety and the bladder

Poor Lola would have had this dream. This story shows how a cat's nerves go straight to its bladder. Lola's owner, Julia, called me in some distress as a last resort after exhausting all other avenues of help. She had two cats, Lola and her female companion, Issie, and they were five and three years old respectively. Julia was out at work all day and she had originally adopted Lola and her brother Simon from a local rescue centre. Sadly Simon had been killed on the road outside her house and she had acquired Issie as company for Lola shortly afterwards. Julia was adamant that the relationship between the two females was amicable although she admitted they rarely spent time in each other's company. Lola preferred the comparative security of a bedroom, preferably under the bed, and Issie would lie on her favourite step halfway down the stairs or in her bed in the kitchen.

Julia had been deeply distressed by the loss of Simon and she became wary of providing her cats with unlimited access to outdoors. When Issie arrived she decided to block up the cat flap and allow both the cats out only under supervision. Lola had only gone out for short periods previously and Julia was convinced she only went outside to relieve herself in the flower bed. Issie was only a kitten when she arrived and Julia felt she wouldn't miss the freedom. Julia dutifully provided them with soft beds, good food and a litter tray discreetly positioned in the corner of the kitchen.

Harmony reigned until Issie was about two years old. Lola had never really explored her maternal feelings towards Issie and appeared to find her invisible or irritating depending on her mood. Issie's increasing maturity seemed to coincide with a widening gulf between the two and little if any interaction. It's amazing how this lack of relationship had never really been appreciated before Julia and I talked. She had always presumed they were friendly and good company for each other because they didn't fight. When she actually thought about it she realized they were rarely in the same room.

Shortly before Issie's second birthday Julia returned home to find a present on the front door mat. One of the cats had passed a flood of strong-smelling urine on the coconut matting. Julia was instantly of the opinion that this was an accident and, although a little surprised, she cleaned up and presumed that was the end of it. Unfortunately this was not the case and over a fourteen-month period there were a number of incidents in various places. Sometimes the quantity of urine passed was small, sometimes large. After a relatively short period Julia realized the culprit was Lola and she took her to her veterinary surgeon to make sure she wasn't ill. Her vet gave her a course of antibiotics in case it was an infection and for a while the problem seemed to resolve. Unfortunately a pattern started to develop with Lola peeing where she shouldn't for a couple of weeks and then returning to normal use of the litter facilities or outdoors. Julia could not understand why she should do this and even started to resent her. After all, she managed to pass her bowel motions in the right place every time, so why was it so difficult to wee there as well? Julia was given a host of well-meaning advice from 'rubbing her nose in it' (I really don't know why this piece of nonsense is still around) to changing the litter material, to putting down tin foil, lemon juice, vinegar and pine cones. All of these suggestions had no effect whatsoever apart from making Julia miserable and frustrated. Lola had started over-grooming her tummy and her skin was pink and sore-looking. She also developed a weight problem. Julia found herself living with an overweight bald cat that she didn't even like any more!

When we think about a cat suffering from cystitis we tend to think of one that passes small amounts of bloodstained urine whilst in intense pain. It is rather obvious for even the most in-experienced owner to spot and usually such cases respond well to treatment. However, other types of bladder disease can occur and one recently identified problem has its origins in stress. It is particularly prevalent in overweight sedentary cats with limited access to outside who live in multi-cat households. Lola was beginning to fit the bill.

Issie was making a status challenge against Lola. She was all grown up and definitely felt she should be in charge of her rather ineffectual companion. By carefully positioning herself on the stairs or in the kitchen she could easily block Lola's access to the one acceptable toilet site in the house. Whilst the contents of your bowel can be held for a significant period, retaining urine is uncomfortable and even painful and eventu-ally the need to pass it becomes urgent. (Bear in mind this stuff dissolves walls!) What better way to assert authority than to dictate when your adversary can use the toilet?

Poor Lola's every waking moment was ruled by access to the litter tray. Persistent urine retention can damage the lining of the bladder and she was suffering from bouts of discomfort which caused her to urinate in any of several inappropriate locations and also led to her over-grooming around her stomach in response to the pain. Julia was mortified that Lola had lived under such dreadful control from the little cat that had originally been purchased as a companion to keep her happy.

With the help of Lola's vet we immediately took action with the appropriate diet and medication for her condition. It was important that she increased her fluid intake so she was placed on a wet tinned diet with high water content. If I ever treat a house-soiling problem I have a very important rule for calculating the number of litter trays required. It is essential to provide 'one tray per cat plus one' in different locations throughout the house to put an end to the possibility of litter tray guarding once and for all. In households with eleven cats this is obviously inconvenient but it does make sense. It won't mean that each cat will use one tray exclusively but it increases a housebound cat's freedom of choice about safe places to pee, and stress levels are immediately reduced.

I gave Julia various pieces of advice about stimulation for Lola and a strict calorie-controlled diet to reduce her weight. Julia was delighted with the response and Lola started to behave more like her old self. Issie was probably gutted. I shall never know.

Smokey – sudden-onset fear

Whilst Joey, Jessie and Lola responded as befitted their character and personality, not all fearful cats are born that way. One particular case I saw some years ago illustrates this point. I remember a telephone call and a rather desperate voice saying, 'Can you help me, please? My poor cat has developed a fear of the colour black!'

It is amazing that some telephone calls can still get me really excited as I suspect the discovery of a new and challenging condition. A potentially melanophobic cat seemed interesting until I thought for a moment and decided further information was necessary. The owner was very worried so I arranged to visit her the following week.

Teresa and Paul were a lovely couple, devoted to their

three-year-old blue Persian called Smokey. This cat had enjoyed an idyllic existence prior to the onset of his current problem. He roamed freely outside when Teresa and Paul were at work but always returned at six o'clock for a fuss and a cuddle. He was a normal confident cat until, one fateful day, something went badly wrong. Teresa found Smokey very distressed when she returned from work and she rushed him to the vet immediately. After various tests were performed nothing was diagnosed apart from a rather inflamed throat. He returned home on medication and made a gradual recovery over the next few days.

About a week after he first became ill, Paul remembered that Smokey behaved quite alarmingly at the sight of a large black dustbin bag that Paul was carrying as he entered the room. Smokey leapt visibly into the air as all the fur on his back and tail stood on end. Since then Smokey had retreated to a small area in the living room, venturing out only very rarely and with great trepidation. This bizarre jumping reaction had been repeated since in response to a black handbag, a black telephone and a very dark navy dressing gown. I was starting to understand how the owners came to the 'fear of black' conclusion.

I have learnt several important lessons about the art of cat behaviour counselling. Always find out what people actually see their pet doing rather than how they interpret it. I realized that Smokey was obviously experiencing some crisis of confidence but I had always doubted the role of colour in his apparent fear. Cats can see in colour but they are colour-blind to red, which appears black to them. They should however be able to perceive blues, yellows and greens. Cats would probably consider our appreciation of colour rather self-indulgent since it seems to be of little significance to them.

Smokey's reaction to the items was more likely to be due to their relative novelty. He was on red alert, seeing danger everywhere for some reason, and any object waved in front of him to test the 'black' theory was bound to provoke a response. As human beings we have an innate need to make sense of everything but I soon learnt that it was pointless to fly around in ever decreasing circles to try to find the reason for *all* unusual behaviour. Sometimes cases like Smokey's remain a mystery; what happened to him that day? Was the sore throat relevant? Despite the fact that my job is 75 per cent detective work I felt justified in merely trying to fix this problem rather than finding the original reason.

When I visited poor Smokey it was quite evident that he was very vigilant. He was comparatively comfortable in the small area where he felt secure but as soon as I moved him away from his corner he was braced for action. Outside his safe haven he actually leapt in the air in response to a red bag, a green box, a white pillow and basically anything novel that was placed in front of him. The 'black thing' had been a red herring. Luckily the owners were resigned to never knowing the reason why their cat had suddenly become fearful and were happy to do anything to restore him to his previous relaxed self.

We tackled the problem by gradually exposing him to a wider area, room by room, and offering the reward of company, affection and food treats to give him a pleasant experience. Almost instantly we saw an improvement. The previously confident Smokey appeared to catch on very quickly that there was really no reason to be anxious. We removed all novel objects from the floor, the odd discarded sports bag or briefcase, just to make sure we didn't 'spook' him again. He continued to make progress and is now fully recovered.

I have often seen this apparent loss of confidence and fear

reaction and it seems to occur in previously confident cats in response to a particularly traumatic experience. With careful thought and patience the problem can be overcome. Smokey was lucky, with no after-effects from his bad experience. Persistently shy and nervous cats remain a worry. I am often asked to see these shy cats with a view to improving their quality of life and making them less fearful. Whilst there are several suggestions that can always be made with regard to managing their problem, the task is equivalent to turning a genuinely shy and nervous person into a confident public speaker. Not easy! In cats who have been timid from birth the best that can be hoped for will be a cat that responds fairly well to the family members but shies away from contact with new people or experiences.

Tips for coping with an anxious cat

- Do not stare at the cat since direct eye contact is challenging in their language.
- Try wherever possible to stick to a routine and remember that any environmental changes are potentially challenging and frightening.
- Try to encourage your cat to indulge in the natural behaviour of hunting by using fishing rod toys or other objects that simulate prey. Even the most nervous cat can lose himself in play occasionally.
- If the cat has a good appetite or enjoys a particular food treat try to encourage him or her to eat it from your hand to increase the human/cat bond. This should be done in moderation to ensure you do not end up with a fat cat!
- Allow the cat to seek out hiding places where he or she feels safe and try not to disturb the cat there.
- Do not actively try to get him or her to face new things by

enforcing your will. Exposure to new experiences should appear to be the cat's choice rather than yours. (Based on the principle that if you want to encourage your cat to frequent a particular room, try shutting him out of it!)

- Understand that interaction with this cat may not be in the usual tactile way and could involve instead gentle kind words or even games.

Flower essences

Many owners over the years have telephoned me for general advice about their nervous or shy cats. They don't want a consultation, just reassurance that their cats are as happy as they can be given their tendency to jump at the slightest noise or run at the first sign of trouble. It is very hard for me to give encouragement without knowing the specifics but there are a lot of cats that behave anxiously and probably always will to one degree or another. I go through the general tips that I've outlined above but often mention the possible use of flower or herbal remedies (such as skullcap and valerian) as a gentle and safe aid in the 'treatment' of anxiety.

Flower remedies have been around for years and I personally have been using them for patients since 1988. When I worked with the RSPCA we often gave a drop of Bach's Rescue Remedy to injured wildlife prior to treatment. I was always surprised at the difference in recovery between those that had received the drop and those that hadn't. Since then I have, from time to time, recommended the use of specific remedies to work alongside behaviour therapy. I must admit that I have no real scientific evidence that they work (homoeopathy is a bit of a mystery to me) but plenty of anecdotal evidence that they can really make a difference.

There is a wide variety of remedies that can be used and each targets a particular negative emotion that your pet may be experiencing. The flower essences are predominantly for treating humans but there are several very good books available that give details of dilutions and treatment suggestions for a range of domestic animals. I always recommend that you check with your veterinary surgeon first before administering any herbal or homoeopathic remedy. The flower essences are preserved in a grape alcohol and if your cat is taking certain medication, such as metronidazole, it may induce vomiting. Otherwise I have found them to be perfectly safe. Here are the remedies that I have used to treat fear and anxiety.

Aspen   This remedy is useful for treating those cats like Spooky that were probably born nervous. It's also very good for those jumpy ones that panic when something new or unexpected happens. I've also used aspen for cats that are frightened of going outside or wet themselves when bullied by another cat.

Larch   This remedy is very useful for those cats that lack confidence or are easily intimidated by other cats.

Mimulus   This remedy is good to treat those cats that have a very specific fear such as other cats or car journeys. It's also very good for the generally timid cat.

Rock Rose   This remedy forms part of the combination treatment called Rescue Remedy (which incidentally is great to take just before driving tests, exams or speaking in public). Let's hope you never have to use this one because it is recommended for those cats that are so terrified that they could harm themselves in trying to escape the danger.

Walnut   This remedy is not specifically for anxiety but it is great for treating those cats that find it hard to adjust to new circumstances or environments. Timid cats love routine and any changes to it can affect them profoundly.

When you have found the remedies that you feel best suit your cat's emotion it is important to dilute them before administration. Two drops of each remedy chosen should be added to a thirty-millilitre container of spring water. This solution will remain fresh for up to three weeks, and four drops should be given four times a day. It can be placed in food or water or placed directly on the tongue in the more compliant patient! It may just make life a little easier for your 'scaredy cat'.

Sadly, owning a timid cat can be very frustrating and practical considerations such as visits to the vet and medication can often be a nightmare. Hours of patience and gentle attention can sometimes reap rewards but avoiding the kitten or cat that appears fearful or hides when you first view it is probably the better option.

❧   ❧   ❧

Spooky lived with us in four different homes over the next ten years, until the day when she was peacefully put to sleep in my arms on the settee in our cottage in Cornwall. Veterinary examination and X-rays indicated that she had damage to the hinge joint on her jaw and the conclusion was that she may have suffered a blow to her face when she was young which had caused her problems in later life. Maybe poor Spooky was one of those fearful cats that had genuinely been cruelly treated. Over a period of months and despite every possible treatment her mouth sealed shut due to arthritic changes in the joint and there was nothing we could do to make her better.

However, I never regretted the decision to adopt her, and the love that she showed to my husband and me was worth every second of patience and frustration.

CHAPTER 3

The Aggressive Cat

Bln's Story

MY HUSBAND, PETER, HAD BECOME TOTALLY HOOKED ON CATS but he did find Spooky rather frustrating at times and felt that he wanted a cat of his own. We returned to the cattery where we had first seen Spooky to find her a companion. It was bursting with prospective candidates but we were both keen for the new cat to choose us and not vice versa, so they were let out one by one to enable us to gauge their reactions. When his turn came, one of two bright orange kittens shot out of his pen and immediately launched himself at Peter. It really was love at first sight and the beginning of a very intense and loving partnership between the two. We made our decision and went home to get our cat basket. When we returned and walked back towards the cattery we could hear loud and insistent miaowing from the pens. When we looked in, the chosen kitten was clinging to the top of the bars and staring towards the door. Peter was convinced that the little kitten was frantic that his new-found friend wouldn't come back for him. We left our basket, with the door open, on the floor of the pen as we started to talk to the cattery supervisor about our new addition. Before we knew it the kitten was in the basket, curled up and fast asleep, awaiting transport to his new home.

I can't quite remember why Peter called him 'Bln'. I think it has something to do with a secret childhood language and a balloon, but who knows. It is pronounced just like 'balloon' without the vowels and always used to cause some confusion when introducing him to people. There are very few cat names that are unique but I am sure this is one of them. I'm pleased that this is the case because he became the most extraordinary companion for Peter. In the early days he was a real handful and he never seemed to stop playing and rushing around the house. Spooky and Bln, surprisingly, adored each other but it was clear that she would rather love him from a reasonable distance. He was far too excitable for her.

As Bln matured, his bond with Peter increased and he spent most of his time with his paws either side of Peter's neck, licking his face all over. I must admit his fishy breath would probably have dissuaded me from such close contact but Peter didn't seem to notice. He used to play very roughly with Bln from the day he arrived home and his constantly lacerated hands and arms looked more like a self-harming problem than the handiwork of a loved pet. Bln soon learned the delights of playing in a dramatically violent way and that habit remained throughout his life.

I would never describe Bln as an aggressive cat since that is a word with malevolent connotations; that being said, he would frequently draw blood and 'attack' both Peter and myself. It is, in any case, unfair to label a cat as aggressive without fully understanding the reasons for the behaviour. For example, many apparently 'aggressive' cats are motivated by fear, pain or disease. Others are merely displaying an element of their natural behaviour, particularly with regard to inter-cat aggression and predation. The fact that we find it unpleasant doesn't make it evil or necessarily wrong. There is, however, another type of 'aggression' that some cats display for which we, as owners, are totally responsible!

Monty – play and aggression

When I first met Monty he was extraordinarily like Bln at the same age and his problem behaviour rang a very familiar bell. Monty was fourteen weeks old when his owner, Isobel, called me and said that she was very worried about her kitten's extreme aggression. Looking back I feel I may have appeared a little patronizing because I couldn't imagine a tiny kitten being remotely dangerous. However, I accepted that she had become truly frightened of Monty and I arranged to visit her and her partner, Andrew, the following week.

When I arrived Monty had been secured in the kitchen and I was shown into the spacious and very stylish living room. It had beautiful laminate flooring and over-stuffed sofas and was remarkably devoid of cat paraphernalia. I then proceeded to take a history from Isobel to get some background before meeting 'the beast'. Monty had been obtained from a domestic home at the age of approximately seven weeks. Isobel and

Andrew were both working full-time and had no other pets. Monty was an indoor cat since their apartment was on the second floor of a large Georgian building.

Apparently during the first two weeks in his new home he was very withdrawn and would growl or hiss at Isobel if she came anywhere near him. With love and patience he soon improved and became more and more confident, investigating strangers and novel items with interest. The problem started after he had been in the flat for a few weeks. Monty started to pounce on passing human legs and lash out with his claws during play. Andrew, being used to dogs, had played physical games with Monty since his arrival. These games involved 'rough and tumbles' using his hand to simulate a sibling's play fighting. Monty enjoyed these interactive games and appeared to have incredible energy. A few weeks prior to the consultation he had escalated his scratching behaviour to biting and would launch himself, not only on Andrew and Isobel, but also on anyone who arrived at the apartment. He would run across the furniture at great speed, straight over any humans who happened to be sitting there. Isobel appeared to be the major target for this behaviour. She had been badly scratched on several occasions and, as a result, was extremely reactive when attacked. She openly admitted that she would scream and cover her face with her hands or fling her arms around to try to deter him. Needless to say it didn't.

Armed with this information I felt it was time to meet Monty. Isobel seemed very anxious at this point and deeply concerned that I would be attacked. I remember making a slightly facetious remark like, 'Go on, let the tiger out of the cage!' and sitting back casually to show her how brave I was. As she opened the door there burst forth a blur of ginger that appeared to have some sort of foot spin problem as he tried desperately to get some purchase on the laminate floor. The running on the spot soon created the necessary momentum and he galloped directly towards me. As he became airborne I sensed the irony of the smug grin on my face and braced myself for the impact. Well, Monty proceeded to bite my hand, my arm, my leg, my notepad, my pen and anything that looked as if it might taste good. As the blood poured I tried to maintain my composure and dignity by writing (as best I could) some extremely relevant comments on my rapidly shredding pad. Those bloodstained notes are now preserved for ever behind glass and they hang on the wall in my office. If ever I get complacent or cocky I just need to look at that trophy of my complete stupidity and it works wonders. Let's face it, the lady said it hurt and I really should have listened.

All this happened in the space of seconds but it had a dream-like timelessness and seemed to go on for ever. When the dust eventually settled and Monty moved on to more responsive targets I took the opportunity to tell Isobel and Andrew why I felt Monty behaved so outrageously. He was an active and confident kitten, I explained, with a considerable need for stimulation and he had learnt (from his physical games with Andrew) to use his human companions as substitute siblings for play fighting. Boisterous games were tremendous fun, just at the time in his development when other kittens in his litter would be helping him learn to regulate his biting by either biting back or interrupting the game. What he was getting instead was a reaction from his owners that he found intensely exciting, which further reinforced his desire to continue the behaviour. Attention from Andrew and Isobel was rewarding and stimulating, therefore any activity that got a response, for example attacking or running all over people, became learnt behaviour and part of everyday life.

Punishment is ineffective in cases of this kind. Monty had directed most of his aggressive play towards Isobel because she was the most responsive recipient. If she had tried to reprimand him it would merely have been another element in an exciting game. The slippery laminate floors were also contributing to the problem. Monty could move around the room far more efficiently if he travelled across furniture and people. This gave the unfortunate illusion that he was performing a 'wall-of-death' fairground stunt and merely added to his owners' theory that he was uncontrollably mad.

Monty was not an incurable lunatic. There are very effective ways to deal with this sort of problem by encouraging the culprit to 'unlearn' bad behaviour. Monty was provided with exciting alternative activities and his owners were instructed to 'ignore' inappropriate behaviour. A comprehensive therapy programme was created to reform this feline felon.

His food was changed gradually to a dry complete veterinary formula for kittens. This enabled us to place it around the apartment to allow him to 'forage' for his food. As he became used to the concept, the acquisition of the food became more challenging as we put it inside cardboard boxes and paper bags. This is a technique that I use a great deal when I need to provide indoor cats with a stimulating environment. Hunting and consuming food would naturally take a cat a great deal longer than a ten-second visit to a bowl in the kitchen. Using dry food to simulate the whole hunting experience is far easier than secreting small amounts of smelly tinned food that can fester quite unpleasantly under the sideboard.

I encouraged Isobel and Andrew to play alternative games at the times when Monty was traditionally more active, using a toy attached to a rod and string that could be agitated in front of him to encourage a reaction. This allowed the game to be totally remote from the owner's body. They had been using a similar toy on a string but Monty's eyes were straying upwards and spying the much more attractive hand attached to it. A long stick or rod makes this process almost impossible and required him to focus totally on the toy. Andrew built a complicated network of cardboard boxes, paper bags and tubes that could be explored and knocked over to provide further excitement.

I asked Isobel and Andrew to ignore, as best they could, the aggressive biting and scratching and the wall-of-death stunts. They were asked not to make eye contact, move or show any emotion. Andrew adapted to this extremely quickly but Isobel was apprehensive. How could she possibly ignore such violence? After all, she was getting hurt. It was eventually agreed that she would only be able to ignore the behaviour if she was suitably protected with strong clothing. Poor Isobel agreed, much to Andrew's delight, to go about her normal business at home dressed in leather boots and trousers with a biker jacket and gauntlets. When she became concerned that Monty could possibly target her face in the absence of other visible flesh I decided to go the whole way and include a full-face motorcycle helmet to complete the ensemble. Isobel was now happy, if somewhat hot.

I asked Andrew and Isobel to give praise and attention to Monty when he was quiet and resting. This would involve stroking but only of short duration. Mainly he was encouraged to sit with them on the sofa. The results were extremely encouraging and Isobel learnt to relax. Monty was praised for acceptable behaviour and he learnt extremely quickly that the biting and scratching wasn't nearly as much fun without the screaming and flailing arms. He loved playing with his new toys and searching for his food and he developed into an active cat who was 'non-violent' towards humans. When Monty was about a year old Isobel and Andrew moved to a large house with a secure garden and Monty discovered the delights of a natural outdoor life. He became quite an assertive character in the neighbourhood and defended his garden with the 'minimum of force', according to his owners.

Emotionally incompetent neutered male cats

Bln had an equally effective technique to defend his territory that kept many a cat at bay over the years. He would square up to a neighbourhood interloper and then scream in the most blood-curdling way. It worked every time; the opponent, suitably wrong-footed, would retreat and no blood was spilt. Bln didn't walk (bless his heart); he minced. Despite his effeminate tendencies he did occasionally become the top cat as our feline group expanded, but more assertive and confident individuals often deposed him. Let's face it, my lovely ginger pussycat would not cope well if he was thrust into the harsh realities of a wildcat existence.

A large percentage of all our neutered male cats would get a dose of the vapours if confronted by a serious aggressor with the intention to kill or be killed. I truly believe that if their lives depended on it they would fight to the death but, if lucky enough to survive, be in therapy for ever. It's a testosterone thing. Tomcats with balls are brave, territorial, strong, motivated, mean and ruthless towards other male cats. They develop thick skin around their necks and strong jowls. Their muscle-bound bodies are built to fight and their pee could stink for Britain it is so laden with male pheromones. They live for violence and sex. However, if you take away that testosterone, preferably before they get any ideas about their maleness, what are you left with? A bunch of soppy mummy's boys! Neutered males are affectionate and soft and we all think they love us and would rather spend time with us than concentrate on being a cat. The reality is they are terrified of going outside because next-door's cat looked at them again with that really horrid expression and it's raining and a mean new Burmese has moved in over the road and there was this loud bang the other day and Tigger from number three is on the prowl and . . . In other words, pretty much the ideal home-loving and friendly cat for us, but not the best emotional state for them.

Hercules – territorial aggression

The territorial thing is relevant to this story because there is a breed that seems to have ignored the fact that we took away their male hormone. Yes, it is the Jekyll and Hyde of the cat world: *the Burmese*! Now before you shift that cute brown creature with the heart-shaped face onto your other shoulder and cry 'Never!' listen to this cautionary tale.

Hercules (Herk for short) was a handsome four-year-old blue Burmese. He lived with his owners Ted and Angela in a very respectable and affluent area of north London. He was a loving pet towards his humans and perfect in every way apart from one. As soon as he ventured out of the cat flap he became a thug and a bully and systematically beat up and terrorized all the cats in the neighbourhood. Ted and Angela were blissfully unaware of his behaviour until a rather embarrassing encounter at a neighbour's barbecue when the residents of the entire street realized that their cats were all being victimized and bitten by one monster and *that* belonged to Ted and

Angela. One particularly sad tale related to an elderly couple, Sam and Doreen, with an equally elderly female cat. Herk had, allegedly, 'broken in' to their home one day through a fanlight window and attacked little Sukie, the cat, whilst she lay sleeping peacefully in her bed. Sam tried desperately to intervene and save his beloved pet and he got badly bitten as Herk lashed out at anything and everything that stood in his way. Sukie had required a spell in hospital and extensive veterinary treatment and she was still recuperating. Doreen felt Sukie had never been the same since and would probably never feel safe in her home again. Ted and Angela were mortified and, in the true tradition of the parent, in complete denial that this could be their wonderful Herk. They called for my assistance.

If I had known then what I know now I would not have got involved. Angela was pleading with me on the telephone so I agreed to visit but at least had the sense not to make any promises. I spent some time talking to the couple and playing with a gorgeous and gentle blue Burmese called Herk. There was definitely a split-personality thing going on and I could see the owners' confusion. I explained to them that this phenomenon is something I see regularly in my caseload. Burmese are renowned for extreme territoriality and aggression towards other cats. Many adult cats find the presence of others in close proximity very challenging and often stressful. Burmese are equally uptight about other cats but they tend to favour a more proactive approach. They take an almost sadistic pleasure in seeking out adversaries, often in their own homes, to beat them up and claim further territory and 'dens' as their own. If people try to defend their victims, the Burmese have no hesitation in redesigning them too with teeth and claws. They are so notorious for their behaviour that they even have nicknames in the behaviour counselling profession of 'despots', 'thugs' and 'bully boys'. Almost without exception they remain incredibly affectionate and gentle pets at home but their raison d'être outside is violence. There are very few things that can be done to deter them. Increased activity indoors and periods of confinement are helpful but not curative. The responsibility often falls on the victims' owners to defend their pets. Owners are not liable if their cats trespass or cause damage resulting from normal feline behaviour. Legally, however, there is a potential liability if damage is caused to property or person by a cat that is allowed to roam when the owners know it is extremely aggressive. The difficulty there is proving what degree of aggression in cats is natural. Ted and Angela were alarmed by this news but very willing to follow any advice concerning measures they could take to lessen the problem.

After about an hour Angela asked if I would go with them to visit Sam and Doreen, as an arbitrator, to show that every effort was being made to resolve this problem. I agreed. Never again!

Sam and Doreen were charming and they sat me down in a comfy chair and offered Belgian chocolate biscuits and freshly ground coffee. I met Sukie, a pathetic little creature, with a plastic cone round her head to prevent her from tampering with a drain that was poking out of a large wound on her side. I sympathized with the view that this was the handiwork of a psychopathic cat. Unfortunately Sam and Doreen's friendly veneer crumbled as I discussed with them various ways they could protect Sukie from future assaults. Shutting the windows, blocking up the cat flap, supervised visits to the garden, aversive techniques to deter Herk were some of the many suggestions I made that were all greeted with a very emphatic 'No'. Sam and Doreen's air changed and the 'genial host' attitude turned into genuine hostility. I was definitely out of my depth and drowning rapidly. Dear old gentlemanly Sam became threatening and abusive and, for the first time ever, I found myself scared of a pensioner. Whilst backing towards the door I explained that I had tremendous sympathy with their frustration and I was, after all, only trying to help. Ted, Angela and I escaped and I went on my way suggesting that they have a chat about my suggestions and give me a call the following day.

The call didn't come and I merely felt that they were taking longer to deliberate their plan for the future. I had suggested that they confine Herk temporarily as things had become so heated. I was therefore shocked to receive a tearful phone call from Angela to say that Ted had been involved in a fistfight with another neighbour. They had answered a knock at the door late one evening and an altercation had taken place concerning vet bills for a bite wound apparently caused by Herk. Unfortunately Ted had misunderstood the timing of the assault and thought that he had uncovered a con artist at work. After all, Herk was tucked up in bed and had been under house arrest for several days. So arms flailed and eyes were blackened and Ted spent his first ever night in a prison cell. All for the love of a cat!

We did try very hard but when the anonymous death threats started to arrive through the mail I felt it was time to take a reality check. Herk could not be permanently confined; it was distressing him greatly. He was therefore allowed out and he immediately resumed his battle strategies. He had to go.

There are always going to be occasions where the environment is just not right for the cat being treated. Herk was highly motivated to defend his territory, to the point that he was perfectly willing to employ pre-emptive strikes. Whilst this is probably an extremely successful technique it is in direct conflict with normal domestic life. Our own cats live in close proximity to many others in a typical residential setting and we expect them, like us, to make every effort to get on with the locals. It just isn't acceptable for us to break in next door to beat up our neighbour over a 'right of way' dispute. It is equally untenable for our cats to behave in this way. In Herk's case it just wasn't feasible to continue. He was simply a 'square cat in a round house' and there was no way he was ever going to fit in.

We set to work to find him an alternative lifestyle and setting that was more suitable for his needs. Fortunately Angela had a maiden aunt who lived in a wild and rural location in Cumbria and we turned to her for help. She had a cosy end of terrace cottage in a sleepy village surrounded by fields and woodland. There were few other cats in the area apart from a small colony of farm cats and the lady was perfectly willing to give Herk a try. A tearful but resigned goodbye took place and Herk moved to his new home in Cumbria. He settled in well and adjusted to his new 'hunting, shooting, fishing' lifestyle. He would occasionally disappear for a couple of days and return home exhausted, maybe with a torn ear, but apparently incredibly content. As far as I know, he is still no particular problem for the local population and their pets.

Tips for coping with a territorial cat

- Encourage your cat to stay indoors at night.
- He or she might find a tasty late night treat sufficient incentive to come in by a certain time.
- If your cat spends time outside during the day it may be useful to shut the cat flap at night and inform the neighbours so that they know when their cats are safe.

- If your cat is nocturnal in his or her habits it may be useful to shut the cat flap during the day and inform the neighbours accordingly.
- Ensure there are sufficient warm beds around the house to give your cat every opportunity for relaxation in a comfortable setting.
- Provide plenty of stimulation indoors (active play sessions etc.) to use up energy.
- Suggest the neighbours have exclusive entry system magnetic cat flaps and ensure that several cats from different households do not possess the same 'keys' on their collars.
- Territorial cats should have a couple of bells attached to their collars so neighbours and their cats can hear them coming!
- Neighbours should be encouraged to keep a water pistol by the back door. One squirt and an element of surprise may deter the less determined cat from entering a neighbour's house.
- Always appear to be doing your very best to resolve the problem. After all, it could have been you and *your* cat on the receiving end!

Aggressive behaviour is a fundamental element of feline life. Let's face it, a little kitty wouldn't last long in a natural environment if he tried to use reason with other cats in territorial disputes and lived off vegetation because he couldn't bear the thought of killing anything. It is almost inevitable that aggression will rear its ugly head at some point, even in the lives of our most placid cat companions.

A small nota bene to all those devoted Burmese owners. Most Burmese are absolutely adorable and the breed in general is probably one of the finest. Every now and then they go wrong, in very small numbers, but when they do it really is dreadful. Sorry!

Tallulah – redirected aggression

Poor Tallulah's story is a good illustration of a type of aggression that occurs when somebody is just in the wrong place at the wrong time. Tallulah's owner, Mrs Stubbens, was a lovely and very elegant elderly lady living alone with her two gorgeous Persians. Tallulah and Tobias were brother and sister and, in their owner's estimation, perfectly happy and content indoor cats. They had lived with Mrs Stubbens for nearly seven years without a moment's trouble. Tallulah had her routines and rituals and so did Tobias. These rituals rarely included any pleasant interaction between the two cats but Mrs Stubbens felt they loved each other as much as she loved them (just in their own way).

Sadly, a week before she contacted me, the most dreadfully unexpected thing had happened. Mrs Stubbens was not very steady on her feet and whilst adjusting a picture on the wall she stumbled backwards slightly. Unfortunately she hadn't noticed Tallulah who had been sitting directly behind her. The next bit happened in a split second and was extremely shocking. Mrs Stubbens trod firmly on Tallulah's tail causing a scream of pain. Tobias happened to be walking through to the kitchen at the time for his mid-morning snack, inadvertently very much (as I said) in the wrong place at the wrong time. Just as Mrs Stubbens's foot came down on her tail, Tallulah launched herself on Tobias in a counter-attack. Poor Tobias got completely trounced by Tallulah in a blur of fur, saliva, teeth and claws. Mrs Stubbens tried to intervene to separate them (never do this at home!) and got badly scratched too. Eventually Tobias escaped to a secret hideaway and Tallulah was left, breathless and dishevelled, looking slightly confused.

Several cups of tea later Mrs Stubbens went in search of Tobias to give some comfort and any necessary first aid. He was fortunately only suffering from superficial injuries but he was extremely shaken. Mrs Stubbens decided to take him downstairs to enable Tallulah to apologize for her actions and to return the household to its normal state of peace. Fat chance! Tallulah expanded to twice her size immediately on seeing Tobias and screamed like a banshee. Her brother stiffened instantly in Mrs Stubbens's arms as he plunged his claws deep into her chest. Mrs Stubbens then screamed like a banshee herself and Tobias flew out of her arms to seek refuge, for the second time that day, in his secret hideaway. He *definitely* wasn't coming out again, no matter how much his owner pleaded!

Tallulah recovered her composure and continued to wash her bottom. Mrs Stubbens got the antiseptic cream to tend to her own injuries and realized that something very bad was happening. One week later both cats could still not be in the same room. I arranged to see her as a matter of urgency.

After hearing her dramatic tale I tried to explain to Mrs Stubbens what had happened. Tallulah had experienced a sudden pain when her owner accidentally trod on her tail. Her instinctive response was to defend herself against a perceived attack and she focused on Tobias as the nearest moving target. Cats have a very keen survival instinct that utilizes an internal system referred to as the fight/flight mechanism. Adrenalin is released and the muscles are pumped full of blood to prepare the body for danger. This massive emotional response was so novel and probably so intensely rewarding that Tallulah became aroused every time she saw the original stimulus for that aggression (poor innocent Tobias). There wasn't anything else going on in Tallulah's life that was nearly so exciting. Cases like this of redirected aggression are relatively common. Unfortunately, if there is no particular love lost between the two pugilists in the first place, a single event such as this can constitute 'irreconcilable differences' and the pair have to be separated for ever. I was worried that Tallulah and Tobias would never see eye to eye again.

I created a therapy programme that included increased activity indoors to try to divert Tallulah with more acceptable pastimes. Mrs Stubbens's stylish London property benefited from a walled courtyard and I asked her if she would consider an enclosed area to give her cats an exciting 'playroom' outside. These outdoor pens, however small, can supply much needed sensory stimulation for the bored housebound cat and I felt this could be the key to distracting Tallulah from her newly discovered vicious streak.

Plans were drawn but over the next few weeks I became worried that things were not going terribly well. Poor Mrs Stubbens was suffering from frayed nerves. The pressure of living on a battlefield, together with the need to follow certain important instructions, was taking its toll. Tobias and Tallulah continued to respond negatively towards each other and Tobias soon adopted a strategy of aggression too to fend off his insane sister. Mrs Stubbens found it impossible to remain calm and the whole situation turned into a horrible nightmare. She couldn't get to grips with building an enclosure; she was struggling with even the most basic functions and the prospect of organizing a builder was daunting to say the least. Eventually, a broken woman, Mrs Stubbens decided to permanently divide her home into two distinct areas and start a new life. Tallulah and Tobias became 'single' cats with one devoted owner who probably to this day is still splitting herself in two to ensure her perfect (?) Persians are happy. Cats, eh?

Rover – the assertively aggressive cat

I have seen many interesting relationships develop between cat and man but not all are quite so innocent as dear Bln and Peter's. Some cats want the total compliance that Peter offered Bln but seem unable to achieve it without using sinister and violent methods. This is a hard concept to embrace but John and his old cat Rover illustrate the point well. John was a gentleman in his fifties who lived alone with Rover in a ramshackle and untidy house. Over the years the two had acquired very similar personalities, gruff, short-tempered and unsociable. John worked from home and he would often focus totally at his desk and completely forget to feed Rover. The wise old cat soon learnt that a smack round the ankles with a full set of claws and a loud hiss would break his concentration and the meal would be provided. They both shared the same favourite chair facing the garden and many a time there would be a battle of wills as either Rover or John defended this seat with rigid determination.

The cantankerous pair rumbled along until John met Gordon. Gordon was a handsome young man, half John's age, and it was apparently love at first sight for both of them. It wasn't long before Gordon moved into John's home and, to Rover's absolute astonishment, his bed. Rover had snored and scratched at the end of John's bed for ten years and suddenly he was usurped. Rover commenced battle and

I was soon called in to prevent further bloodshed.

Gordon was a delightful young man and deeply concerned about the problem with Rover. He understood John's sense of divided loyalty and desperately wanted Rover to like him. Unfortunately the furious old cat attacked Gordon on sight and herded him round the house like a malevolent collie. Rover had certainly won the first battle in the war; Gordon had moved into the spare bedroom. I spoke at some length to Gordon whilst John popped in and out of the room making contributory but not very helpful comments. For example: 'If Gordon just stood up to him everything would be OK.' However, having seen the state of Gordon's lacerated legs and his obvious terror when Rover patrolled past the doorway, I was sceptical that Gordon could stand up to anything just then.

Once again Rover's behaviour had developed over the years on a trial and error basis. As far as he was concerned the best way to get what you want was by tooth and claw. He had tried the plaintive miaow and the gentle rubbing round the legs but this had proved a waste of time. As the theory of learning states, if you are rewarded for an action you are likely to repeat it. So when Rover first became frustrated and lashed out at his owner the instant response pretty much guaranteed he would do it again. There are many cats like Rover throughout the country and I have certainly met a few over the years. They are at their most successful when they are presented with a gentle, compliant and responsive human. John was a little distracted most of the time and Rover only used his aggressive requests in certain circumstances. Gordon, however, was a completely different story and Rover soon saw that he could get him to jump through hoops by just looking at him. The power! The possibilities! So Rover shifted into overdrive and started to control Gordon completely.

When I suspect I will be encountering an assertively aggressive cat like Rover I have a determined mantra that I mumble on the way to the consultation. 'It's only a cat, I'm bigger than him. It's only a cat, I'm bigger than him . . .' You get the picture. By the time I arrive at the house my air of superiority and dismissal is at its peak and it works every time. The cat is immediately taken aback at the obvious lack of response and his (or her) demeanour changes to one of intense curiosity. Basically I ignore the cat. Aggressive posturing and spitting is only powerful if your victim is paying any attention. Otherwise you just look stupid.

Whilst this technique is great in principle it is extremely difficult to ask someone who has spent hours bleeding in casualty to ignore his assailant. Gordon was not convinced this would work. Unfortunately, as I was making some closing comments prior to departure, a rather unfortunate shift of power took place when Rover suddenly entered the room in a remarkably menacing fashion. His eyes were wide and black and he was emitting a low rumble. My previous authoritative manner was starting to fizzle a bit and I began to feel very threatened. In my defence, fear is very contagious and Gordon was quivering in a heap. I swear Rover had seemed to metamorphose into something akin to the beast of Bodmin, with the biggest teeth I had ever seen. I suddenly realized that I was trapped. The consultation was coming to an end, the programme had been discussed and agreed upon and I knew there was no way that Rover was going to let me leave the room. Bugger.

As always, Vicky had a Plan B. Never let it be said that she would ever appear to have lost control! I started, once again, to go over the details of the therapy programme and allowed Gordon to believe that I just hadn't finished talking. I reiterated the following points. Rover should have a change of feeding regime from two distinct meals a day to a more 'ad lib' arrangement of feeding bowls in several locations within the home. This would avoid that dangerous time when Rover was hungry and Gordon just wasn't quick enough. John and Rover's favourite chair was moved from its current location facing the garden and placed in John's study. Gordon was often attacked if he passed Rover lying on this chair so, if the chair was removed, there was little point in just standing there. I also suggested that it was perfectly acceptable for John to shut Rover out of the bedroom at night and provide him with a warm thermal blanket on the spare bed as a reasonable alternative. Gordon was instructed to wear suitable protective boots and leggings whenever he was in the house. I explained that there was no way that Rover (despite the fact that his teeth really were huge!) could penetrate these defences and, providing he wore the boots, Gordon would not be attacked again. I went through the 'ignoring' technique and told him how to focus straight ahead and walk with confidence in Rover's presence. Ironically he seemed reassured at a time when I was losing my cool big time. I continued to talk for quite a while, long after incurring a £40 penalty on my parking meter. I tried desperately to ignore Rover and I constantly requested that Gordon maintain eye contact with me. Eventually John was our saviour. He casually booted Rover up the bottom as he entered the room and the tension was diffused. I saw my chance and escaped quickly.

I was resigned to the fact that Gordon would find it extremely difficult to get the upper hand with Rover but obviously underestimated his determination and his delight in having an excuse to wear leather jackboots round the house. After several weeks I received a beautifully handwritten note to say that, whilst not exactly bosom buddies, Rover and Gordon had appeared to reach a truce. Rover had not even attempted to attack Gordon for at least two weeks and he appeared highly satisfied with the new feeding regime that enabled him to eat whenever he felt like it. He continued to hiss occasionally if they passed in a narrow corridor and Gordon stated in his letter that he felt he would have to wear the boots for a significant period to avoid any risk of further aggression. Exactly my sentiments. Well done, Gordon!

Predatory 'aggression'

There are many different motivations for aggressive behaviour in our pet cats but probably the most natural is their response to small prey and their incredible hunting skills. Bln was a great hunter in his home in Cornwall, an area rich in mice, rats, rabbits, birds, voles, shrews, lizards, slowworms (need I go on?). He was an efficient despatcher and, with great consideration for my sensitivities, consumed everything of the tiny prey apart from the gall bladder. Over the years I have had numerous telephone calls, all with the theme, 'How can I stop my cat bringing back dead animals?' I'm afraid my answer is usually, as politely as possible, 'Why would you want to stop a natural behaviour?' 'Tiddles the predator' is always a slightly uncomfortable concept for many people but this is fundamentally the nature of the beast. If you are having a problem with corpses in your kitchen, and really cannot bear it, then there are a few suggestions that you may find useful.

- It may help to confine your cat indoors at those times when his or her hunting trophies are at their most abundant. This may be the case during dawn and dusk and at night. Many cat charities recommend that cats are kept in at night in urban areas so this is a fairly sensible routine to adopt anyway. If your cat is not used to staying indoors then build up the length of confinement gradually. For example, one hour the first night increasing to two hours by the third night and so on. When your cat is indoors it is important to offer plenty of play and games to compensate, as well as warm beds and quiet resting places. Playing with furry toys and feathers on sticks will not make your cat a more proficient hunter.

- Try adding two small bells to your cat's collar. These may knock together and alert any potential prey to the danger. Many cats get wise to this very quickly and hold their neck so still a large cowbell wouldn't sound if it were dangling there!

- Ultrasonic collars are available that emit a high-pitched sound to warn prey but these also are not 100 per cent effective.

- Even if you are a bird lover it is probably best not to encourage birds into your garden with feeders and bird tables. If you feel compelled to do so then it is advisable to make the stand of any table as high as possible.

- If a small rodent is running round your kitchen then ensure you attempt its capture using an old oven or gardening glove. They bite!

- Better still: get someone else to catch it (if you can find anyone).

- Worm your cat regularly. Prey animals are often hosts for certain parasites and this can lead to a revolting burden that could become evident from either end of your cat. Sorry about that!

- Do not feed your cat more food to stop him hunting. The urge to hunt is not motivated by hunger.
- Do not punish your cat for bringing in prey, dead or alive. He or she is merely doing what comes naturally and choosing to bring food back to the den.
- Watch your fishponds outside and protect them with netting. Your cat may be into fishing, too.

Muffin – pain-induced aggression

I have come across many instances over the years where an understanding of pain has been essential to my work. Pain can often affect a cat's behaviour profoundly and previously placid and amiable creatures can become claw-flailing monsters.

I have seen hundreds of cats and their owners over the years but I can always manage to remember something that takes me back to their home in my head. I have so many mental maps in my brain that once again I marvel at its capacity for storing information. Muffin's owner Sally sticks in my mind because she had an amazing collection of pastel-coloured china rabbits. They were all produced in the 1930s, if I recall correctly, and the manufacturer is extremely collectable apparently. I remember regressing slightly at the sight of them since a large blue bunny featured heavily in my childhood, sitting on the hearth in my mother's family home. Apart from a collection of blue and green rabbits, Sally also had Muffin and her companion, Petal.

Muffin was a five-year-old, rather rotund tabby and white moggy. Her companion, viewed by Muffin as a deeply resented interloper, was three years old and a dear little long-haired tortoiseshell. Petal came and went through the cat flap, ate her food and went upstairs to bed. She was a bit of a lodger but occasionally she would reward her owner with some affection-ate contact. Muffin, however, was very different. She hung round Sally like a lovesick teenager and she cried for a cuddle, she cried for food and she even cried to be picked up and put on the sofa because her little legs 'just couldn't really do the job too easily these days'. Muffin ate a diet of expensive tinned cat food (too much, unfortunately) and loved her owner. That was pretty much her life. She didn't go outdoors very often (could she actually get through the cat flap?) and Sally felt that she was frightened of the other cats outside. She also thought she was a little apprehensive of Petal.

There wasn't much room in Sally's pretty little cottage (what with the rabbits and everything) but she still managed to provide everything she felt her cats would need. They had a litter tray in the bathroom upstairs, several cat beds dotted around, an upright scratching post, a few toys and all the food and love they could take. What more could a cat want? Well, obviously something, since it was evident that Muffin had a problem.

The soiling had started about three years previously. Muffin had started to wee on the carpets in the dining room and living room. Every time it happened Sally vowed to do something about it, but just as she was about to make the phone call the problem stopped. Sally was after all a busy lady and she managed the problem as best she could until the situation deteriorated shortly before my visit. The strong smell of urine as I walked through her front door certainly indicated that Muffin was in the middle of 'one of her episodes'. Sally couldn't even smell the urine any more so I figured this was a bigger problem than she even realized. Once again I found myself in a rather embarrassing posture in a stranger's house.

Little did my mother know that my expensive education would result in my spending my life on all fours sniffing people's carpets. The strong stench of ammonia was in various locations around the edges of the carpets in both the living room and the dining room. The urine had undoubtedly soaked through the carpet to the underlay and concrete beneath. Shampooing the carpet was by now a pretty pointless exercise.

The reason why Sally had called for my services at that point was the sinister twist that had taken place. Muffin's whole personality had suddenly changed. She had transmogrified into a devil cat, definitely with evil on her mind. Sally could no longer pick her up without receiving a tirade of spitting abuse. Worse still, her four-year-old niece had been the victim of an un-provoked claw attack across the face. Sally felt the poor child would probably now be phobic for life and she felt extremely guilty. What on earth had happened to the lovely gentle Muffin she used to know?

The answer soon becomes apparent if you recall the case of Lola and Issie and the big litter tray issues. Muffin was suffer-ing from a stress-related urinary tract problem and her personality had changed because she was in leg-crossing, tummy-wrenching pain! She hated going to the toilet outside, she hated Petal, she hated sharing the litter tray and she fluctuated between periods when she coped and periods when she didn't. Sally had recently changed her job and she was working longer hours. The shift in routine was too much to bear and this particular episode of bladder problems was a real humdinger. Every cat's response to a similar sensation will be different. Lola over-groomed in response to the pain and Muffin tried to take a child's eye out. The coping strategy may be different but in both cases there was obviously something very wrong in their world that needed addressing.

I then started to do the detective work to try to discover what was bothering Muffin. I had already identified her poor relationship with Petal and her apparent distaste about sharing a litter tray. I also knew she was reluctant to explore outside (let alone pee there) because of other cats. It soon became clear that Sally had fallen completely out of love with her so we had an important relationship to restore as well as everything else.

The veterinary surgeon agreed totally with my thoughts on the urinary tract problem and prescribed a supplement that helps to repair long-standing damage to the bladder. I gave Sally advice about increased activities for Muffin including aerobic exercise (yes, cats need it too) where Sally could be involved. So every evening, with a glass of Chardonnay in one hand and a fishing rod in the other, Sally entertained Muffin with a single feather on the end of a piece of string. I also made sure there were plenty of easily accessible high places and secret hiding areas where Muffin could feel safe from her enemies. The litter tray was an enormously significant problem for Muffin so, with the tried and tested formula of 'one tray per cat plus one' we found two further private locations for a couple of brand new and highly desirable cat toilets. The carpet was beyond redemption so Sally had no alternative but to take it all up together with the underlay and destroy it. Any remaining odour of urine in any private corner of the dining room or living room could potentially have fooled Muffin into presuming it was an acceptable toilet. Sally was highly delighted to see the back of the garishly patterned carpet chosen by the previous owners (maybe Muffin was just making a gesture for good taste). Sally had always promised herself laminate flooring and, once the problem was sorted, she said she would treat herself.

The results were brilliant. Muffin, the devil cat, returned to the depths of Hades and the old friendly, rather rotund tabby and white cat was back. The extra litter trays were a big hit and Muffin immediately christened both the new trays and continued to do so regularly. Sally eventually got her beech laminate and Muffin agreed to disagree with Petal. What a wonderful result, but a great illustration of how pain can cause aggressive behaviour.

❁   ❁   ❁

Just before his fourteenth birthday, my own 'aggressive' cat, Bln, became ill. It was obvious to me that he was losing weight and that his heart was racing. He had also started to howl at night. I suspected that he was suffering from hyperthyroidism (a tumour on the thyroid gland) and took him to the vet. This condition is relatively common in older cats and it causes major metabolic changes that put an enormous strain on the heart. My suspicion was confirmed and we tried to medicate him with tablets prior to surgery. Unfortunately, with our very best efforts, this proved virtually impossible and Bln's health started to fail. When Bln was operated on, the vet was unable to find a thyroid tumour. She suspected that it was ectopic tissue near the base of the heart that was causing the problem (basically a tumour occurring where it cannot be operated on). Bln was brought round from the surgery and came home but he never really recovered. Shortly afterwards Bln's liver failed and he was put to sleep on 13 March, the very same day that we had lost Spooky nine years previously.

CHAPTER 4

The Indoor Cat

Bakewell's Story

SHORTLY BEFORE WE WERE DUE TO MOVE HOUSE, AN OLD friend called from Derbyshire to say that a local farmer was about to drown some kittens for whom he had failed to find homes. The word had soon got around that Vicky had transformed from a hard-nosed businesswoman to a mad cat lady. Our friend obviously knew that I wouldn't let this heinous act take place but would intervene to rescue the kits from the brink of death. I checked with the local RSPCA cattery where I was working at the time and phoned the friend back to say, 'Bring them down to Kent and we'll find homes for them here.' It was late that evening when our friend arrived, so we had to keep the kittens overnight before transporting them to the cattery. Two beautiful black ones (about four months old, one male and one female) jumped out of the cat basket to explore their new surroundings. They were both extremely gentle and friendly little souls who showed no ill effects from their long journey. We kept them in the utility room overnight and when I opened the door in the morning they were both sitting upright in the bed and looking expectantly towards the door. I swear there were two tiny little halos above their heads and they oozed sweetness and light. They had used the litter tray and eaten their food and were obviously keen to look like perfect pets. Well, needless to say, they never found their way into the cattery. Number three and number four cats in the original group, they integrated extremely well. We called them Bakewell and Tart originally (after the pastry associated with their home town) but when we found out that it was called 'Bakewell pudding' in Derbyshire we were greatly relieved. Bakewell became a loving and gentle cat and Pudding (Puddy) became my absolute darling, the love of my life.

Shortly before we moved from Kent to Cornwall we found ourselves temporarily homeless. We had ended up completing on our Kent home quickly as a condition of the sale and we had nowhere to go. My brother Stephen stepped in and suggested we stay with him in his small house in Biggin Hill until the purchase of our next property was completed. We were there for seven weeks. Stephen is not a great animal lover and the prospect of sharing his home with four cats and their litter trays appalled him. The cats obviously sensed this and, as cats do, saved all their eliminatory habits for those moments when Stephen was eating. He was not amused and to this day I still cannot smell pungent air fresheners without thinking of him. Eventually he insisted that the cats were kept in our bedroom when he was at home and this caused tremendous problems, especially for Bakewell. Spooky, Bln and Puddy resigned themselves to a life of inactivity and slept practically all the time. Bakewell, still only seven or eight months old, was a very active kitten. He really couldn't see the point of just lying there so he set about looking for alternative pastimes to fill the void. Unfortunately, he discovered the delights of DIY and he stripped wallpaper, shredded carpet and consumed an entire window sill and frame (the wood was slightly rotten and great to chew). It was a very expensive seven weeks for us as we eventually had to replace the window and completely re-decorate the room before we felt we could ever look Stephen in the eye again. Bakewell taught me a big lesson during those few weeks. Keeping a cat exclusively indoors without active stimulation is potentially a problem. The whole experience was the origin of one of the expressions that I use most often in my consultations: 'The devil makes work for idle paws.'

Keeping cats indoors

The domestic cat is an animal at its best in a territory rich in suitable prey and challenging landscape, and there is no deny-ing that it is not always possible to offer the perfect home for a pet cat. Stephen's small house in Biggin Hill, for instance, was certainly not ideal for Bakewell. A property on a busy street or four floors up is not conducive to a free-ranging life either, but many owners feel they can still offer their cat a fulfilling existence. Do we really appreciate all the implications of confinement?

We humans have a very big world. We travel, we go to work, we socialize with friends; every day we experience a wealth of different stimuli in an ever changing environment. We return home full of 'input' from the day and curl up with our cat, who has been eagerly awaiting our return. The outdoor cat will have faced similar challenges in the feline version of a typical day at the office: hunting, foraging, climbing, exploring, social encounters. The prospect of a meal and a warm lap lures our companion home to spend a restful evening. But what sort of day has the indoor cat had? Got up, ate, slept, looked out of the window, slept, ate, used the litter tray, slept a bit more, went upstairs, looked out of the window, slept, ate, used the litter tray again, slept, owner came home. Or, in Bakewell's case, ate a window sill. Where is the hunting? Where is the climbing? Where is the challenge?

The indoor cat has a very small world. Everything remains the same apart from the comings and goings of the owners and the cat's own movement about the house. There are plenty of places to sleep and occasionally good vantage points from which to view the outside, so most cats, like Spooky, Bln and Puddy, in the absence of other more interesting distractions, will sleep to pass the day. They will not be run over by a car, a neighbour's cat won't bite them and they will not be locked in someone's garage. We have guaranteed our indoor cats a physically safe environment but it is also our duty to ensure we provide for their mental health. Here is a tale that illustrates all too vividly how profoundly some cats can be affected.

Molly – the 'stir crazy' cat

A young girl called Lydia lived alone in a one-bedroom flat with her companion cat, Molly. They had lived together for four years. Molly never went out so was always there when

Lydia came home after a long day at the office. Lydia felt Molly was perfectly happy. She had a nice bed, loads of food and a pile of soft toys placed in the middle of the living room every day. One morning Lydia woke to see Molly, all fluffed up, staring out of the bedroom window and growling in a blood-curdling way. Lydia was shocked. She had never seen Molly like this, so she leant forward to touch her and give her comfort. Suddenly Molly exploded and viciously attacked Lydia with teeth and claws and then rushed round the flat, growling and hissing. Lydia was petrified and she pushed Molly into the kitchen with a pillow and shut the door firmly. For the following two days she lived under siege. Every time she opened the door Molly was ready to pounce on her, eyes wide and ears flat. Lydia called the vet who referred her to me. I arranged for a thorough physical examination (during which Molly behaved impeccably) and found that she was perfectly healthy.

In the absence of a clinical cause for this extreme behaviour it was clear that we were dealing with an even more distressing problem. When I visited Lydia I was intrigued by Molly's apparent good behaviour when the vet examined her so I was keen to test her response to me. I entered the kitchen and awaited an assault, armoured with a stout pair of boots underneath my trouser suit as always. Molly certainly looked frightened and highly aroused but she made no attempt to attack me. However, when I coaxed her into the living room, where Lydia was perched surrounded with a barricade of cushions and pillows, she immediately fired up and started spitting and growling. I tried to explain to Lydia why her beloved pet had become her enemy.

Poor Molly had experienced an emotional overload. She had probably seen another cat outside and become incredibly aroused by the sight. Her body prepared her for a fight with an enormous rush of adrenalin and Lydia's touch caused an explosion of aggression that represented four years of restricted behaviour. She had finally gone 'stir crazy' and was unable to calm herself. She then became locked into a cycle of dysfunctional behaviour, experiencing the same extreme emotional response every time she saw Lydia. Other people could handle her without any problems, so Molly was rehomed immediately to an experienced 'foster parent' and given a chance to recover and live a fuller, more active life. Whilst it cannot be denied that this is an unusual case, there are many other stress responses that a cat can exhibit when it is unable to behave naturally.

Activity budgets – assessing your cat's lifestyle

Cats are designed to live a particular kind of life and most free-ranging felines follow a similar pattern of behaviour. This natural pattern can be described via an observational tool called an activity budget. In this particular context the word budget relates to a unit of time and not money and it describes the proportion of a twenty-four-hour period that a cat spends performing certain natural behaviours. All cats have an activity budget that is dependent on their lifestyle (indoor, outdoor), breed, character, age and external influences. A cat in a natural state, who is allowed to exercise freedom of choice, can spend for example fifteen hours sleeping or resting, three hours grooming and playing and six hours hunting, eating and exploring. Why don't you look at your own cat's activity budget just to see how it compares with a natural lifestyle? It is when your cat's activity budget starts to differ dramatically from natural behaviour that you potentially have a problem.

I tend to use a budget with six different categories: sleep, interaction with surroundings, time spent outside, social interaction with people or other cats, grooming and eating. I must admit I did scare Bln, my feisty red tabby, when I started research into the lifestyle of the domestic cat and was exploring the concept of 'activity budgets'. I diligently equipped myself with a pen, clipboard, stopwatch and plenty of caffeine and observed Bln closely for a whole day and night. As I watched him, Bln sensibly slept for eighteen hours but every time he woke he was most distressed to see a hunched creature with bloodshot eyes staring at him from the corner of the room. Our relationship became distinctly cool for several weeks afterwards. I learnt a lot from that experiment, not least that you really shouldn't stare at your cat for prolonged periods because it makes him paranoid.

Just to give you an idea, here is Bln's original activity budget from 1994 and one relating to Puddy in her twelfth year:

Bln's activity budget

Bln was six years old at the time and part of a group of seven cats. He was lucky enough to enjoy a rural location with free access outdoors. However, it was a very wet day and Bln hated getting his feet muddy. He was provided with indoor litter facilities, an enriched environment within the cottage and unlimited access to Peter.

Sleep	*17.5 hrs (very boring part to monitor)*
Environmental	
Interaction	*3.6 hrs (looking out of windows mainly)*
Outside	*1.1 hrs (not on the muddy bits)*
Social interaction	*1.3 hrs (licking Peter's face or smacking Puddy)*
Grooming	*0.2 hrs (a very lazy grooming regime; face only)*

Eating   *0.3 hrs (cat food or anything from the fridge)*

Puddy's activity budget

Puddy was twelve years old when I checked her activity budget and part of the same seven-cat group. Same location but she was monitored on a dry but cold day in January. She also had unrestricted access outside and all the other benefits that Bln had.

Sleep	*17 hrs (slept a lot when she was older)*
Environmental	
Interaction	*1.5 hrs (looking out of the window)*
Outside	*1.5 hrs (mainly staying near the house)*
Social interaction	*3 hrs (following me around)*
Grooming	*0.5 hrs (quite a thorough job)*
Eating	*0.5 hrs (another cat that hung around fridges)*

So, we are trying to justify keeping cats indoors, and rather than saying 'sure, they look perfectly happy' we should really be assessing them in a more scientific way that is more sympathetic to their species and their natural behaviour. Pet ownership is not just about giving pleasure to the owners. It really should be a two-way thing and it is worth looking at what the cat gets out of it in a confined environment.

Immediately, you can see from the activity budgets of Bln and Puddy that there is a chunk of it that cannot be fulfilled by the indoor cat. Time cannot be spent outside and (if the cat is being kept on its own with owners out at work full-time) social interaction is extremely limited. This spare time has to be filled with something; it cannot just be a void. The cat then has a limited choice of ways to spend that time and the easiest option is through sleep. In the absence of environmental stimulation it is unlikely to fill its time with interaction with the environment so it's left with eating, sleeping and grooming. Now can you see how cats that are kept in unstimulating environments indoors become obese and sedentary and can develop a tendency to over-groom?

This is a terrible oversimplification of a complex subject but I am sure that you are now beginning to realize that it probably isn't the best option to keep cats exclusively indoors. If, for whatever reason, it is necessary to do so it is essential that the right cat is chosen and that the environment is as stimulating as possible.

Stimulating the indoor cat

A territory that can be viewed in its entirety from one location can never provide sufficient interest for a cat, so people who live in bedsits or one-bedroom flats should think again about owning a cat if they cannot provide access to outdoors. Any property with more space may be suitable but care should be taken to provide the cat with a little more than just food, a bed and a litter tray. Commercially available cat toys are great fun but, sitting in a pile on the floor, offer little if any incentive for the cat to chase or pounce because they are not moving. Some cats will make their own games by tossing these 'mice' into the air but this sort of activity is best left to periods of shared quality time between owner and pet.

If the owner is out at work during the day then there will be many hours for the cat to fill, so other non-interactive provisions need to be made. It is important to remember that the principle of 'environmental enrichment' is to enable the cat to perform natural behaviours in an artificial setting. Hunting, foraging, climbing and exploring can all be enjoyed with a few cardboard boxes and a bit of imagination. Dried food can be placed around the house in various locations, as we did for Monty in Chapter 3, to allow the cat to work to satisfy its hunger rather than visiting the predictable bowl in the kitchen. When you consider how much effort and mental energy goes into hunting, killing and consuming one mouse it is possible to understand how unrewarding mealtimes become for a house-bound cat. Climbing is very important and opportunities to create a three-dimensional space will increase the area that the cat has to live in. Wardrobes and tall cupboards and shelves can offer attractive viewing platforms and resting places and these should be made accessible wherever possible. Cardboard boxes and paper bags are always worthy of exploration providing they contain a little catnip or dried cat treat. Sleep takes up a large proportion of a cat's time even in the natural environment, so plenty of secluded and private resting places should also be available.

It can never be an ideal situation to keep a cat exclusively indoors. The nature of the cat is such that it is at its best when exercising freedom of choice and any restriction on its territory automatically prevents this. However, if it is necessary to have a house cat then here are a few golden rules.

Golden rules for indoor cats

- Obtain a cat or two kittens that have always lived indoors. Whilst the transition from indoor life to outdoor life can be made, it is unfair to restrict a previously free-ranging cat.
- Research breeds beforehand and try to choose one renowned for its placid and docile nature. These cats will require less 'input' and will be less labour intensive to keep amused.

- Provide suitable litter facilities in a private place away from the feeding area, and clean them daily.
- Provide a good supply of food and water. Cats naturally would not seek to satisfy thirst and hunger in the same location. It is essential that cats drink as much as possible, particularly whilst on a dry diet, therefore providing bowls or glasses of water in alternative locations will often make it more attractive.
- The calorific requirements of an indoor cat will be smaller than for a more active free-roaming one. Food quantities should be adjusted accordingly to avoid obesity. It is not a good idea to compensate for lack of stimulation by feeding.
- A scratching post should be positioned so that the cat can use it at full stretch. Many of the commercially available posts are designed for kittens and tend to be rather short. Several posts around the house are ideal and might prevent the risk of damage to the sofa. Tall 'cataerobic' centres can be purchased and, if positioned correctly, will provide a useful and popular play area.
- A small amount of grass should be provided since cats need access to vegetation to aid digestion. Tubs of grass seed are produced specifically for this purpose.
- Care should be taken about having houseplants since some can be poisonous to cats.
- The cat's safety should be considered, particularly with regard to the kitchen and various appliances.
- Do not assume that two cats will automatically be 'company for each other'. If the property is large enough to enable each cat to have its own space, then it may certainly work very well. However, some cats resent the presence of others and the answer may be for the owners to spend more quality time with them.

- Every opportunity must be taken to stimulate the cat. Grooming, stroking and games should all be offered if the cat enjoys the activity.
- Fresh air is important and windows should be opened slightly for a period every day or lattice screens fitted that allow the window to be opened more fully.

Behavioural problems in the indoor cat

Whatever lifestyle your own cat enjoys, the ability to see things from a feline perspective will always ensure that you are doing your very best. Cats have numerous talents including adaptability and this is one of the many reasons why they are so popular. However, there are times when a bored cat can easily become focused on issues that would not normally be important and these can create stress. For example, two adult cats together may cause one to focus too intensely on the antagonism and sense of competition between them. A cat may become over-attached to its owner or 'obsessed' with food. The most worrying result could be that the poor cat performs repetitive sequences of behaviour that herald the onset of a stereotypical or obsessive-compulsive problem. Just think of the pacing tigers and the weaving bears in the dreadful old-style zoos and you will understand what I am saying. Many of the behavioural cases I see involve cats kept exclusively indoors. Their paws are inevitably idle unless preventative steps are taken.

Ying and Yang – the destructive Siamese

Two cats that I remember with some amusement were a couple of active male Siamese with very idle paws indeed. Ying and Yang were elegant creatures living with their owners, Claire and Steve, in a second-floor flat. Claire called me in some distress to say that she had come to the end of her tether. She had tried hard to cope with a long-term problem, following advice from friends and the Internet, but she was now admitting defeat. I promised I would see what I could do.

Ying and Yang were very different characters but, physically, like peas in a pod. Identification, however, was easy by observing their response to me as I entered their home. Ying immediately approached me, sniffed me with great curiosity, and then leapt on my shoulder and proceeded to 'groom' my hair using teeth and claws. I had been warned by Claire of his tendency to do this to newcomers so I was prepared. I was actually quite relieved, because she had also said that if Ying didn't like the look of you, he would attack you viciously instead! Whilst Ying was restyling my hair into something an apprentice barber would have been ashamed of, I saw Yang at the other end of the room. His attitude to strangers was one of suspicion and he would always take himself off to a safe distance and chew a piece of fabric in an absent-minded fashion. Many cushions, curtains and throws had been destroyed over the years by his determined molars. As I was watching Yang, his brother decided that he had given me enough of a beehive to satisfy his artistry and he jumped off my shoulder. He then proceeded to indulge in his second favourite habit, 'making love' to a piece of his owner's clothing. Masturbation in cats is one of those taboo subjects that is occasionally alluded to vaguely by owners with phrases such as 'humping', 'naughties' and 'rumpy pumpy'. The target of the cat's affection is usually either an item of clothing (preferably smelly) or a cuddly toy. Castrated males are the biggest exponents of the art of masturbation and it is often merely a symptom of an underlying problem, such as frustration, boredom, conflict or anxiety, but in the absence of any other coping strategy it can be extremely rewarding. I have occasionally been asked to address this as a specific problem; I was even asked by one lady to train her cat to stop masturbating on her leg but encourage him to suck her ear lobes instead! Needless to say I never got round to seeing that one. In Ying's case it was very much a symptom and, if I recall correctly, I didn't make a big issue of it. I merely sat back and listened to their story whilst gently distracting Ying by offering the more socially acceptable pastime of playing with a feather on a stick.

Claire and Steve had adjusted over the years to an extremely unusual lifestyle. Every evening they came home to a house that had been completely trashed: CDs strewn across the floor, food out of cupboards, lamps off the tables and magazines ripped to shreds. They never gasped in horror, phoned the police or checked to see what had been stolen. Every day they gathered up the CDs, returned the food to the cupboards, put lamps back on tables and threw away magazines. Ying and Yang had entertained themselves over the years by behaving like the most atrocious vandals and no matter what Claire and Steve did, they continued to wreak havoc when their owners were out at work during the day.

After a lengthy discussion, I felt that I was witnessing the very worst example of idle paws. Both cats were highly intelligent individuals with an enormous need for 'input' but their experiences were limited to the static environment within the four walls of their home. They were not climbing, exploring, foraging or having social encounters in an ever changing world outside and they were both frustrated and bored. So they had tried to fulfil the need to perform natural behaviour in their artificial and unchallenging apartment with disastrous results.

Claire and Steve were sceptical that a solution was possible but I was confident that we could make an enormous difference to Ying and Yang. I would however need a great deal of co-operation and enthusiasm from the owners. During my tour of the flat I was shown the third bedroom, which appeared to be a storeroom for boxes and general paraphernalia. I tentatively asked whether or not they would be amenable to a little 'DIY makeover'. I felt the room would make a perfect cat play area that might well offer some of the many challenges of the great outdoors. They agreed and I enlisted the help of a very good friend (far better with a saw and hammer than me!) to transform their junk room into a 'state of the art' cat amusement arcade.

We had several hours of clearing stuff out before we were left with an empty canvas. The first thing we did was pull up the old carpet and secure it to the wall with the aid of double-sided adhesive carpet tape and a couple of wooden batons. Immediately we had an exciting and novel cat climbing frame! Wooden posts, covered with thick sisal twine, were secured to the floor and ceiling to provide a structure that represented everything from a rickety fence to a sturdy tree. A series of shelves were attached to the walls to provide exciting vantage points from which to view their new playground. Several dark cupboards contained warm and cosy beds for those private times of rest. Thick knotted chandler's rope was attached to the ceiling to provide a challenging climbing opportunity. The floor was covered with an assortment of textures to roll on and scratch. Objects such as cardboard tubes and paper bags were dotted everywhere, within which were secreted little meals of their favourite dry biscuits. Receptacles were attached to the carpet on the wall and dangled from the ceiling, providing opportunities to exercise more demanding problem-solving techniques to obtain food. There were cat-friendly plants, attractive litter facilities and indoor water fountains. The room was an indoor cat's paradise as far as I was concerned but it needed to be put to the test.

Their owners brought the two Siamese into the room and Ying immediately attached himself to the wall carpet as if he had a Velcro belly. Claire and Steve were delighted with the results and we all had our fingers firmly crossed when they went to work the following morning. Would the room make that big a difference? Well, believe me, it did. Ying and Yang didn't even leave it for a week and completely ignored any comings and goings in the rest of the flat. With the aid of a little help from medication for Yang's fabric-eating habit he showed no interest whatsoever in the cat bedding in the room.

Claire and Steve were delighted and Ying and Yang ceased their destructive and unsociable habits in favour of their new activity emporium. I was extremely pleased with my achievement until, two years later, I had a telephone call from Claire. She thought I might like to know that they had just decided to get a dog. They had not made the decision lightly but the cats just weren't company any more as they spent most of their time 'doing stuff' in their own room. Oops. Maybe sometimes things can go *too* well.

Harness walking

There is an option that suits a certain type of cat and that is daily walks outside on a harness. Whilst it is not ideal (that whole walk to heel stuff really isn't a feline thing) I have known many cats that have thoroughly enjoyed a saunter in the park in central London or a trip down a leafy lane in suburbia attached to a harness and lead.

In my experience this is an idiosyncrasy common in certain breeds, for example Maine Coon, Siamese, Burmese and Bengal. Good old moggies might follow you for a walk across the meadows but I would warn against trying to attach them to anything whilst they do so! Harness walking may well be a useful antidote for some indoor cats but it is important to maintain the habit since a brief glimpse of the great outdoors followed by persistent confinement may be worse than no exposure at all. Here are a few recommendations if you think this may be an option for you and your indoor cat.

Tips for harness walking
- Ensure the harness is designed for use in cats and it is fitted correctly.
- Start early and get your new kitten used to wearing a harness.
- Attach the harness initially without the lead.
- Reward your kitten and have a game when he/she first wears the harness to distract him/her.
- Attach the harness for short periods initially and increase the time worn gradually.
- Always reward your kitten during periods of harness training with food or play.

- Never leave your kitten or cat unsupervised whilst wearing a harness.
- Bear in mind the harness will need adjusting or replacing as your kitten grows.
- The lead can be attached and allowed to trail once your kitten gets used to wearing a harness.
- Gradually get your kitten used to you holding the lead and following him/her around. Now you are ready to try the process outdoors.
- Ensure the first visit outside is in a safe and quiet area.
- Gradually increase the time spent outside.
- Avoid walking on busy roads or in parks with dogs. Cats can become easily frightened and their instinct will be to escape!

Outdoor enclosures

The final compromise to improve indoor living is the outside enclosure. This does not have to be large or elaborate but it can provide enormous stimulation for the bored house cat, with all the sights and sounds and smells of the great outdoors.

Barnabas – the benefits of an outside enclosure

I met Barnabas and his owner Penny several years ago. Penny was referred to me by her vet with a very distressing problem. She explained to me that, until recently, she had been the proud owner of two male Siamese, Luke and Barnabas. They were brothers and they had remained good companions all their lives. After some delving into the realities of the relationship it was clear that Luke was the more assertive of the two brothers and Barnabas tended to remain very much on the sidelines. They were kept exclusively indoors because Penny lived in a built-up area and she was worried about their being killed on the road or stolen. She never had any problems with them; they were affectionate, playful and great company for her. What more could you want? Sadly, tragedy struck and Luke became ill. He had developed a tumour and he went downhill rapidly until Penny made the painful decision to put him to sleep. He was only eight years old. She was devastated and it made matters worse that Barnabas appeared inconsolable.

Time passed and Barnabas gradually started to change. He became clingy and extremely vocal, crying and following Penny everywhere. She felt dreadful because she knew he was trying to adjust to life without his brother but he was becoming an absolute pain. He cried in the night and he cried in the day. There was no rest for Penny at all and she found herself going out during the day just to get away from Barnabas. When he started to urinate on his bed she realized something had to be done.

I visited her shortly after our telephone call and spent some time getting to know her and the lovely Barnabas. During my visit he was fairly curious about me but his devotion to Penny was incredible to watch. He never took his eyes off her and followed her everywhere, crying and patting her with his paw. Penny, in turn, never took her eyes off Barnabas and her conversation with me was interspersed with frequent asides to the little Siamese. You could tell that she was frustrated by him but she seemed powerless to ignore his constant demands.

The bed-wetting had started about a month prior to my visit. Penny had bought Barnabas a new beanbag for a treat. She thought it looked soft and warm and as if it might encourage Barnabas to have a good night's sleep. Unfortunately all it encouraged him to do was urinate on it repeatedly. Penny was seriously falling out of love with Barnabas. Luke had always been her favourite cat and Barnabas had always been the quiet reserved one. How she wished he would rediscover his quiet and reserved side again!

This was a sad case because I could see why Barnabas was behaving in this way. When two cats live together all their lives it doesn't necessarily mean that their relationship was made in heaven. If one is more assertive than the other the less con- fident individual will often withdraw. They never really display their true personality until such time as the other cat isn't there any more. Barnabas relied on Luke a lot when he was there and he probably felt comfortable with the predictability and the lack of responsibility. Now that Luke was gone poor Barnabas had to make it up as he went along and turned to Penny for moral support and guidance. Penny provided Barnabas with the reward of attention and contact every time he asked and this had created a dependency that was distress- ing both parties (more of this sort of problem in Chapter 7). Stressed cats tend to retain urine and poor Barnabas may well have become confused about the yielding nature of his new bed! I know my darling Puddy used to get terribly confused about the difference between a tray full of litter and a bag full of beans.

I explained to Penny about Barnabas's insecurities and the need to withdraw a little and give the Siamese something to do outside his relationship with his owner. During my con- sultation I had become transfixed by the beauty of Penny's little garden. It was full of beautiful shrubs, trees and bedding plants, with a neat little lawn and patio. She was obviously a keen and accomplished gardener. I started to devise a plan but I had to check with Penny first to gauge her reaction.

Barnabas needed stuff to do. A small enclosure outside seemed to be the answer. The garden was full of sights and sounds and Penny had already mentioned that Barnabas used to love sitting on the window sill looking out at it. If we could encourage him to spend some time away from Penny in a fun environment it might just trigger that change in him that would restore the owner/cat relationship to something more enjoyable. Penny, understandably, didn't want her garden to end up looking like a zoo enclosure so I set about designing a small but stimulating (and aesthetically pleasant) pen with various shelves and levels within it that would provide an interesting place for Barnabas to spend time. Most of all it was safe and Penny didn't have to supervise him or worry that he would come to any harm.

I explained to Penny that I felt the bed-wetting would probably resolve if we threw out the beanbag and provided further trays for Barnabas, including one in his outside enclosure. If we could reduce his anxiety I felt that we would be well on the road to resolving the soiling problem too.

A local carpenter was employed to build the enclosure and he did a great job. It was a wood-framed construction with platforms and shelves and ramps to give Barnabas endless opportunities to scratch and climb and generally lounge around in high places. The roof was slightly pitched and covered with a waterproof material to protect against the elements. Pots of grasses and herbs lined the floor of the pen and there were various areas for shelter from the sun and wind. Penny enhanced the whole effect by including a fountain with trickling water that would entertain Barnabas for hours.

Penny had been told to start to ignore Barnabas's advances

and constant attention-seeking behaviour in preparation for his new and exciting lifestyle. This had created a frustration in Barnabas that merely made him try harder and cry louder, but I had prepared Penny for his Plan B. I had told her that any increase in his demands meant that her tactics were working, and this gave her an even greater resolve to continue. A cat flap had been constructed through the wall of the house so that Barnabas could decide when he wanted to spend time in his new garden room. It was now time to introduce him to the outside world. Penny was delighted with his reaction. For the first hour he merely sat and sniffed the air. Gradually he started to explore and he scratched on the wooden ramps and he chewed on the grass in the pots. He patted at the water feature and he rolled on the ground. Things were looking good!

Over the next few weeks there were various ups and downs. Barnabas still stayed in at night and paced and howled but Penny remained in her bed and determined not to reward his behaviour with her attention. A spell of good weather heralded the start of Barnabas's recovery. He started to spend more and more time outside and even remained there at night on occasions. He did 'talk' a lot still but, as Penny was quick to point out, he is a Siamese!

The soiling problem had been resolved; poor Barnabas obviously wasn't a cat that would see a beanbag as anything but a toilet. The provision of a total of three trays (two indoors, one outside) allowed him to have a choice of toilet locations and the system worked well for him. Penny was delighted and she felt that their relationship had been restored. She still missed Luke but she felt, with time, that Barnabas would become a really good companion.

Tips for outdoor enclosures

- A concrete or patio slab base will enable the enclosure to be used all year round.
- The ideal structure is wood-framed (weatherproofed with animal-friendly preservative) with wire mesh attached to the outside.
- If the enclosure is south facing it is important to provide areas within it that will protect your cat from direct sunlight.
- The roof should always be covered with a waterproof material to guard against the worst of the weather.
- The wall of the house can form part of the structure with or without cat flap access from the house.
- A door should be included in the structure for ease of access and cleaning.
- The structure doesn't have to be large but the full height can be used by including wooden platforms and shelves.
- Shelves can be accessed by logs or ramps made from wood.
- Include pots of grasses or other cat-friendly plants inside the enclosure.
- Include a water feature if there is sufficient room; if not a water bowl should always be available.
- An outdoor covered litter tray may also be provided, particularly if your cat does not have freedom of access into and out of the enclosure.

❋   ❋   ❋

Bakewell certainly taught me many important lessons about the dangers of keeping cats indoors. Over the years he blossomed and became everyone's favourite cat. He can purr for Britain and enjoys nothing more than to lie on your lap with his head buried between your legs. Bakewell grew into a tall, handsome and rangy cat with a long bushy tail and a liquid wobbly walk that is unmistakable. Peter and I always said that if he was a human he would wear beads, eat brown rice and talk about peace and love. He rarely fought in the territory with other cats but preferred to avoid confrontation and keep himself to himself. The only aggression Bakewell ever showed was a bit of swearing and cuffing around the head directed at his sister. Puddy always retaliated with great enthusiasm but they still slept on the bed together.

Bakewell is still with us, the last survivor of the original four, but he took the loss of his three long-term companions rather badly. Life has gone on and Bakewell is now the oldest male and a reluctant leader. It has proved very stressful for him. He even started to spray urine indoors as he desperately tried to cope with the responsibilities of his new position. Many neighbourhood cats, which in the old days wouldn't have dared come near, began to encroach on the previously well-defended territory and hang around our cottage in Cornwall. We had naively believed that the area was devoid of other cats but, in retrospect, the three most assertive cats in the group had done an excellent job in keeping them truly at bay. When they died, Bakewell seemed to grow in physical stature and he became a muscular cat for a thirteen-year-old. Emotionally however he seemed very vulnerable and my heart aches when I see him still trying to make sense of it all. Although he had a typical sibling rivalry with his sister Puddy, it appears that he misses her very much.

Recently Bakewell was diagnosed with hyperthyroidism. He had started calling out in the night and Peter and I were immediately suspicious that this was the problem. Both Hoppy and Bln had done this before their diagnoses and we feared

Bakewell was going the same way. He was given medication for three weeks and then a bilateral thyroidectomy was performed (removal of tumours on both thyroid glands). Unfortunately there was a complication after surgery. The thyroid glands are immediately adjacent to the parathyroids, which are very tiny but extremely important since they regulate the calcium balance in the body. Calcium is vital for just about everything and damage to the parathyroids during surgery can be very dangerous. Within a couple of days after his surgery Bakewell had started to hide on top of the kitchen cupboards. He had a frightened expression on his face and clever Peter realized that he might be suffering from the first signs of hypocalcaemia (low concentration of calcium in the blood). The following few weeks were very fraught with regular blood tests and medication and very worried 'parents'. I was desperate for Bakewell to pull through; he had experienced such a tough time prior to his surgery I just wanted him to have some fun again. I shouldn't have worried. Bakewell may eat brown rice and wear beads and sandals but he's a tough nut! He's fine now.

CHAPTER 5

The Multi-Cat Household

Lucy's Story

LUCY'S STORY BEGINS DURING MY TIME WORKING IN AN RSPCA
centre in Cornwall. One day in May a rather intimidating
leather-clad man on a motorbike brought a little tabby and
white female cat into the centre. He was a member of a local
chapter of Hell's Angels and one of the most unlikely cat
people I had ever met. He told me that he had owned the cat
from a tiny kitten and, from the earliest age, she had gone
everywhere with him zipped up inside his leather jacket.
Unfortunately she was getting a little bit too big and wriggly
and he felt it would be safer to rehome her rather than force
her to fit in with his nomadic high-speed lifestyle. He was a

kind and gentle man and I assured him that his little cat would go to a good home. The best, actually, because I had decided instantly that she should become number five in our cat family. I called her Lucy (soon to be known exclusively as Loose Elastic) and she joined the crowd at home. Lucy is a beautiful little cat who reminds me very much of my first cat, Spooky. Thankfully she has her looks but not her nervousness; she is loving and friendly towards every living thing she encounters. She integrated well and became firm friends with all of the others. At that time I was still naively introducing newcomers with little real understanding of the implications but, luckily, I had chosen a very compatible personality and harmony reigned.

The pitfalls of the multi-cat household

The cases I see in my work as a cat behaviour counsellor make me realize how lucky I was (and still am) with my multi-cat household. I have never really experienced any major problems and I have witnessed genuine companionships between certain members of my cat family. I have tried to look for the chink in the armour of their relationships but I merely see normal feline social interaction with no evidence of anything remotely dysfunctional. Do you think I am viewing them through rose-coloured spectacles? I find myself these days challenging owners of multi-cat households who say that all their cats get on really well with each other, because I am fairly convinced that, if cats were ever interviewed in some parallel universe, the conversation would be thus:

Interviewer Well, Fluffy. Your life must be pretty good. You have a wonderful home and a big garden to play in. OK, your owners are out a lot but you have great company in Tiger, Jasper and Felix. It must be brilliant to rough and tumble and have fun with them all day?

Fluffy   I guess. *(paws folded, sulky expression)*

Interviewer   I understand you have gourmet food on demand *and* freshly cooked chicken daily. That must be great?

Fluffy   . . . S'pose. *(think 'stroppy teenager')*

Interviewer   I cannot imagine you could be happier. So, why are you so depressed? Frankly I think you are being rather ungrateful. Do you realize some cats live with lonely old ladies in draughty cottages in the middle of nowhere, just cat and owner and acres of isolated countryside? How would you like that?

Fluffy   Yes, please! I'll get the cat basket! *(exit stage right at great speed)*

Given the choice, many would do the same, including Tiger, Jasper and Felix. So are we kidding ourselves when we think our cats are such bosom pals?

The social structure of the domestic cat

Let's look at the domestic cat in a natural free-roaming environment without the direct intervention of us pet owners. To say that the cat is a solitary creature per se is an over-simplification. Much of a cat's behaviour is dedicated to its relationship with other individuals but I would argue that this is an adaptive state for the species and a consequence of domestication. Cats are capable of existing in an incredibly flexible and variable group situation of anything from the single cat to as many as two thousand in one square kilometre. These groups are centred around a plentiful food source (usually man-made). On the positive side, social living provides any animal with a defence against predators, a collective resistance to harsh environments and reproductive access to other members of the species. All of these survival techniques are as important to the cat as to any other species.

So what happens when they live with us? We neuter them (usually) and we decide who lives with whom. We then place them within a territory containing, potentially, many other individuals that would not necessarily have chosen to congregate in such close proximity. Do multi-cat households work? Are they beneficial for the cats as well as us? Let's try to answer these questions by looking at the potential problems that may occur.

The 'Feline Felons' survey

Firstly, just to throw a few statistics into the subject, let's look at the results of a survey that was conducted by myself with the help of *All About Cats* magazine. Readers were requested to complete a questionnaire if they felt that they were experiencing behavioural problems with their cats. A total of 267 owners took part, owning between them 784 cats (with one lady sharing herself and her home with thirty furry friends). The questionnaire asked several questions about the household in general, nutrition and lifestyle, and then asked owners to list the various problems that they were experiencing.

The majority of owners (73 per cent) had more than one cat. Seventy-five per cent of the cats included in the survey were domestic 'moggies' and the rest were pedigrees, the top four breeds for problem behaviour in the survey being Siamese, British Shorthair, Persian and Burmese. Just over a quarter of the owners kept their cats indoors permanently whilst the rest allowed them either free or restricted access to outdoors. Of the latter group, nearly two-thirds provided their cats with a cat flap.

Most of the owners who completed the questionnaire were experiencing more than one behavioural problem, either from one particular cat or from several cats in the household. The really frightening statistic was that twenty-seven households were trying to manage eight different behavioural problems at the same time!

This survey provided a great deal of useful information, not least of which related to the potential pitfalls of multi-cat households. Often owners are blissfully unaware of friction between their pets because they judge relationships to be good if there is no fighting. Cats know there are far better campaign strategies for battle! Inter-cat aggression can be passive, subtle and devious but the victim can become extremely distressed by the relentless conflict. Cats have limited ways to express their emotional states so sometimes it is useful to diagnose dis-harmony by default if behavioural problems can be identified within the household. For example, traces of dark sticky urine on your radiator is a good indicator that things are not going according to plan.

Urine spraying

Urine spraying is a perfectly normal feline behaviour and it represents a strong visual and olfactory means of communi-cation within a cat's territory. As cats appear to be able to differentiate between sprayed and squatted urine it is believed that the anal glands may play a part when urine is sprayed to produce the oily, viscous liquid found on our skirting boards. There is theoretically no need to spray urine indoors if it is perceived as the cat's core area. Safety and security in this context should be paramount. If the individual develops a sense of insecurity and becomes stressed, then it has limited ways of expressing this vulnerability. So it uses a natural behaviour (urine spraying) usually employed in situations of conflict. Not all urine spraying is anxiety-related. It may also relate to territorially assertive individuals. In my experience this is rarely the case in indoor spraying problems unless an interloper is coming in from outside.

The jury is still out regarding the definitive reasons why the behaviour is so necessary but it appears to be most relevant to the sprayer itself. Any cat is capable of spraying urine, male, female, neutered or entire. Sexually active cats will spray urine that is laden with pheromones to indicate their readiness for mating. Neutered cats will spray on fences and bushes, for example, in areas of high cat density as part of their daily routine. Despite the fact that urine spraying can be utilized to relieve all sorts of weird and wonderful emotions in certain rather complex individuals it can safely be said that it is usually 'a cat thing' and another feline is at the root of the problem. Now, the issue could be the presence of lots of other frightening cats outside and a sense that they may launch an invasion or, even worse, the aggressor could be the 'enemy within'. Very scary.

Putting my theory about the fragile nature of most multi-cat households to the test was relatively easy using data from the survey. One chilling conclusion soon became apparent. 'The incidence of urine spraying indoors increases in proportion to the number of cats in the home, from 17 per cent of single-cat households to 86 per cent of those with seven or more cats.' Eureka! Every house has that 'one too many' cats threshold. Two may be a crowd in homes containing particularly intolerant individuals whereas seven just may not be heaven in others, but either way a great deal of luck is required – together with the right environment – to prevent all hell breaking loose. Indoor spraying is often just one symptom of a generally turbulent and disruptive existence (only 5 per cent of the households experiencing urine spraying indoors listed this as their only problem). Most urine spraying cases are combined with other complications including excessive scratching within the home, anxious individuals in the group and, worse still, inappropriate urination or defecation indoors.

Floss – bed-wetting in the multi-cat household

Here's a perfect example. A lady called Sue called me to say that she was very worried about one of her four cats. She had suddenly started wetting on her owner's bed after the arrival of Sue's thirtieth birthday present – a one-eared, second-hand cat from a local rescue shelter (personally I'd have preferred flowers). After establishing with her vet that there was no medical cause for this behaviour, I agreed to visit her later that week.

Sue was a bubbly and friendly girl and obviously very eager to understand why Floss, the wetter, had started to behave so outrageously. I sat in her living room and began to write my notes as she started talking under the watchful eye of Sandy, the eldest of her four cats. Floss would probably have joined in also had she not been hiding behind the television in the corner of the room with eyes as round as saucers. Sue's two other cats, Pokemon and Splodge (the aurally challenged newcomer) were elsewhere. As we discussed the background to the problem, a large tabby cat walked into the room with an air of nonchalant authority. I did a quick ear count and said, 'Ah, this must be Pokemon.' Sue replied instantly, 'No, silly, that's Buster. He lives down the road and always comes in every morning for a snooze.' How stupid of me. Rather than voice my reaction at that time, I made some acid comments on my notes. Buster left the room and went out of the cat flap and within minutes the flap clattered again to herald the arrival of Pokemon or Splodge. As a sleek young black cat entered the room I said 'Pokemon?' hesitantly, only to be informed with great pride that it was the little girl from next door called Princess, who liked to come in and play when her owner's children became rather boisterous. I did eventually meet Pokemon and Splodge as Sue diligently dragged one out from under her spare bed and removed the other from a shoe box in her wardrobe. Both returned to their bolt holes as if on elastic when she let them go but at least I'd met them briefly and their chosen locations spoke volumes. We returned once again to the living room and, as yet another new feline face appeared at the window (Claude from the bungalow opposite), I asked the burning question. 'Don't you think it's wrong to have all these other cats coming into your house?' Sue's reply was interesting if somewhat misguided. 'Oh, no. I actively encourage it. I'm delighted my cats have such a good social life!'

Within minutes I had shattered, as gently as I could, all Sue's preconceptions about her cats and cats in general. As I checked each room of her home I discovered a number of other signs that the household's cats were on 'amber alert'. I found areas of scratched wallpaper around the entrance to the master bed-room; small amounts of sprayed urine on the banisters and the skirting boards beside the front door; a very smelly mattress and four stressed cats! Sue was devastated. Occasionally an owner will experience a particularly obvious problem (Floss peeing on the duvet) that is impossible to ignore, and seek help from me, believing this to be the one and only issue in the household. However, this is rarely the case: it is usually discovered to be the tip of the iceberg. All four cats, including Splodge the newcomer, were experiencing a sense of deep insecurity. Anyone and everyone was entering and exiting their 'den' and their home was definitely not their castle. The drawbridge was down and the invading forces were penetrating at will. The arrival of Splodge had merely been the last straw for Floss and her distress had gone straight to her bladder. The others were also in the middle of their own personal nightmare.

I felt that I had a chance to restore some order to the household if I could work exclusively with the four resident cats rather than the entire neighbourhood. Urine spraying and house soiling is often associated with a 'breach in the cat's defences' and the blame is almost always directed towards that most erroneous invention, the cat flap.

This amazing contraption was received with great enthusiasm when it was first launched on the pet market. Great! At last, a chance to give your cat freedom of choice to come and go as he pleases throughout the day when you are out at work. No more days spent outside in the cold; no more days spent indoors with legs crossed because you forgot to go in the flower bed first thing in the morning. It didn't take too long for most cats to get the idea of how to use it and it was heralded as a real breakthrough in pet care. However, I view cat flaps slightly differently. Let me just offer you a scenario in our human world that would represent an equivalent situation.

Your partner returns home one day and announces that he has a brilliant idea. He feels that you both waste so much time fumbling for keys when arriving home at night that he has decided to replace the front door with a swing door that can be pushed with the shoulder to gain immediate entry. No more locks or keys, just 100 per cent freedom of entry and exit. The first night he creates his brilliant swing door you find it rather difficult to sleep. You keep hearing noises and wonder whether it's that new swing door. Is there someone downstairs? You get up several times in the night to peer over the banisters to see if there is someone in the hall. The following day you go out briefly but you find yourself looking round to see if there are any dodgy types about that may just take advantage of your new swing door and let themselves in. You return home early. That evening you are sitting with your partner in the lounge and the swing door clatters and a large and threatening stranger marches into the hall and heads straight for the kitchen. As you watch from behind the sofa he takes food out of your fridge, has a drink and then goes upstairs for a quick lie down before leaving. You are both terrified. That night you take it in turns to guard the new swing door. By the end of the week you are both on the verge of a nervous breakdown as you cover your house with posters of Rottweilers that say 'Go on, make my day', have false alarm system boxes fitted to the walls and put Neighbourhood Watch stickers everywhere. But still they come. The word has got out that you operate an open house and you are powerless to prevent an invasion.

Promise me that next time you see your cat sitting by the cat flap staring at it you will remember the story of the swing door. It may help you to understand what your cat is *really* thinking.

Returning to Sue's dilemma, it was clear that the cat flap had to go. The survey conducted into behavioural problems, whilst not conclusive, certainly gives weight to the argument against cat flaps. Fifty-seven per cent of the households with cat flaps had spraying problems compared with 21 per cent of households with no cat flap. Sue accepted that she had to try to prevent others from coming in but suggested a compromise: she would fit an exclusive entry system with magnetic keys on her own cats' collars to limit use to them alone. Surely this would stop other cats coming in? Whilst the idea is good in theory there are several flaws to these magnetic systems. First, they don't always work because some cats seem unable to hold their necks in such a way as to activate the mechanism. Second, the cats do not understand the principle behind exclusive entry and, as far as they are concerned, it is still a dropped drawbridge. Third, the cats will suddenly become collectors of cutlery and masonry nails as they attract anything metallic they happen to encounter in their travels. Unfortunately there's more. Your cat cannot gain access at a run (this gives them a bad headache as they end up head-butting a solid door) and the thug of a tomcat from down the road can break in anyway with brute force and ignorance. The only thing to do is to remove the cat flap completely to give a strong signal to the resident cats that their defences are once again intact. It is usually sufficient to board up the door (both sides) with plywood to give the impression of a solid door.

Sue followed the instructions and allowed her cats access to the garden on demand. She provided five litter trays in various discreet locations in the house (following the formula one tray per cat plus one) and created new high resting places and private beds. She offered several new areas for food and water bowls to give Floss, Pokemon, Splodge and Sandy an impression of abundant resources. A cat's sense of smell is incredibly sensitive and positive scent messages are a great way to restore a sense of security. Food and catnip can be used together with the cat's own naturally occurring pheromones secreted from glands in their cheeks and around their heads. Research has been conducted into feline facial pheromones and their significance to the average cat and it was discovered that they tend not to spray urine in areas where there is evidence of facial pheromones. A fraction of these pheromones, common to all domestic cats, has now been manufactured synthetically and can be used in homes where urine spraying is a problem. In combination with behaviour therapy these pheromones are a useful tool.

Sue worked hard over the next few weeks and she was amazed at the transformation in her cats. She felt they had become more relaxed and they were tending to play more and rest in more open places. They didn't spend hours mooching around the cat flap area and Splodge was really not that much of a problem for Floss after all. The mattress on Sue's bed unfortunately had to go but things soon became calm and Buster, Princess and Claude had to look elsewhere for their entertainment.

Inappropriate urination

Urination in inappropriate places indoors can be a real problem in multi-cat households and there is a great deal of nonsense advice given to people who are living with this problem. Thirty per cent of households in the 'Feline Felons' survey were constantly mopping up urine and they will know how damaging it is, both to the carpet and the whole owner/cat relationship. Whenever I visit these unfortunate homes (apart from the heady aroma of ammonia reminiscent of old-fashioned perming solutions) I am bombarded with air fresheners of every conceivable kind, tin foil on the carpet, polythene over the sofa, pine cones in the corner, pepper behind the television, orange peel on the stairs and other weird and wonderful 'tricks' which have been recommended to tackle the problem. All to no avail. Deterrents merely redirect the problem to another area because the root cause has not been identified and dealt with. I would strongly recommend you do not waste your time. You are better off limiting the damage by confining your cat to a smaller area until professional help can be sought.

Mitigating circumstances for house soiling

Whilst there is no real substitute for pet behaviour counselling there are always positive things that you can do that may help before calling in the experts. I always tend to think of 'reasons why cats soil in the house' as mitigating circumstances in a courtroom drama. The ideal home for a cat is one where he can plead no mitigating circumstances for the alleged felonious behaviour. Of course, this is really the wrong way to view these things since cats are never bad or naughty when they soil. They are always unhappy because something has gone wrong with their world and they are powerless to put it right without your help. Let's make sure we do everything we can to ensure we are sympathetic to their needs. The list below is not exhaustive, but it goes a small way towards showing you how many little things can become big issues for our cats.

- I'm not well. I have cystitis and I need to wee urgently all the time.
- I'm not well. I have diarrhoea and I can't make it to the tray.
- I'm old and want the comfort of an indoor toilet.
- There is no indoor litter tray and I have been frightened by

a neighbour's cat whilst urinating under the conifers. I'm not doing that again in a hurry.

- My owner has produced a delightful feng shui garden outside but concreted over my favourite latrine area.
- It's pouring with rain and blowing a gale. Why don't you try going to the toilet outside in this weather?
- The indoor litter tray is the same one I had as a kitten. I'm bigger now and I can't turn round and dig in it comfortably.
- The dog stares at me when I'm using the tray. There's just no privacy.
- The litter tray is next to the washing machine that switches itself on suddenly at night. That can be scary.
- The wood pellets in my tray may have been great when I was a kitten but I'm heavier now. It's like walking on Whitstable beach in bare feet. Ouch!
- The polythene liner to my tray gets caught in my claws when I dig and the litter goes in my eyes.
- My owners don't clean my tray out regularly because it's a covered one and they can't smell it. Walking in there makes my eyes water. It stinks!
- The litter tray is so conveniently placed for my owner that it's like an assault course to get there (in the cupboard, over the vacuum cleaner, through the flap, etc.).
- I have to share my tray with another cat. Brother, schmother, I still don't like him and I want a separate toilet.
- My tray is right next to my bowl of food. Disgusting, how unhygienic!
- Sooty won't let me use the litter tray. He says it's his.
- My owners use a pine-scented freshener in my litter. What's pine? It smells ghastly.
- I have to use a litter tray that is right next to the patio doors and next-door's cat pulls faces at me through the window.

- The litter tray is right next to the cat flap. What if a strange cat comes through when I'm using it? I can't defend myself.
- My owners use newspaper to line my tray but object when I pee on the *Sunday Telegraph* before they have read it. Isn't that unreasonable?
- Last time I used the litter tray my owner shoved a pill down my throat. I'm not going there again.
- I'm a Persian, for goodness' sake! What do you expect? *(Sorry, Persian owners! There are many perfectly well behaved Persians but I see far more than I should for this sort of problem.)*

Cleaning soiled areas

Understanding the potential causes of house soiling is half the battle towards finding a solution. Providing the appropriate number of trays indoors (irrespective of the presence of a cat flap) containing a fine grain substrate in various discreet locations will always help these situations. Cleaning previously soiled areas is also essential but most products available are pretty ineffective if the problem has been evident for some time. Cats can potentially pass gallons of urine over a period of months or years and it penetrates the carpet, the underlay and the floorboards or concrete beneath. It is very invasive. There is often no alternative but to remove that section of carpet and replace it. Prior to fitting a new section it is worth treating the wood or concrete beneath to try to remove any remaining odour that may be present. Some litter additives (to reduce the smell from litter trays) contain a mineral called zeolite that is extremely absorbent. This can often collect any remaining gas that may be lurking in the floor if it is sprinkled liberally over the area and vacuumed away forty-eight hours later. Whilst it's working this product looks and smells like an open-plan litter tray so it's important to keep any cats away.

Inter-cat aggression in the multi-cat household

We have now established links between multi-cat households and a variety of stress-related problems. The most common behaviour reported by many multi-cat owners relates to a natural feline behaviour that we find terribly unpleasant. Two-thirds of the owners who completed the behaviour survey complained about territorial aggression and fighting. (This represented at least one cat from nearly all the households that allowed their cats outside.) Some cats are naturally more territorial or aggressive than others; 16 per cent of the cats that fought outside were also aggressive to other cats in the household. As discussed previously, this can be covert or passive as well as active physical aggression. Passive aggression is always more evident since it doesn't make evolutionary sense to fight with such impressive weaponry at every opportunity. Many of the cat's body signals relate to threat or a reluctance to engage in aggressive encounters. Cat owners are often blissfully unaware that their cuddly pets are secretly filled with loathing for their feline companions. Would you recognize the signs? Is inter-cat aggression easy to spot in a multi-cat household? Is it only diagnosed by default (if there is a behavioural problem) or only when there is active fighting?

Here's another interesting statistic that, unintentionally, shed some light on this complicated issue. I conducted an elderly cat survey in 1995 and compiled data relating to 1,236 cats over the age of twelve (more about that in Chapter 8). Part of that survey discussed cats whose cat companions had died. Sixty per cent of those cats (596) who had experienced such a loss had a very noticeable reaction. Lots displayed calling and searching behaviour but many others were reported to have blossomed afterwards and become calmer, friendlier and 'happier' cats. Others, particularly Orientals, seemed to improve with the introduction of a new companion. We are really beginning to see how difficult it is to be sure that all multi-cat households work as well as we think.

Social signalling

Let's look briefly at the cat's social structure and some of the signalling that you are likely to see and should be able to interpret in feline terms.

- A direct stare is challenging and usually employed by more dominant individuals.
- Overt fights are far more likely between individuals if there is no difference in social status.
- High-status/ranking cats may decline to actively fight. They will look away and walk away and sit and groom indicating the other cat has lost. Only really high-ranking cats are able to do this. They are confident that they don't have anything else to prove and they might even spray urine afterwards as a victory display.
- Signalling of a dominant cat can be very subtle and passive and can include anything that elicits a withdrawal or deferential behaviour. For example, standing or sitting in a doorway and blocking access to a desired area.
- Dominant cats control access to litter trays, stay in them longer and use them first.
- Lower ranking males will show their bellies to higher ranking males but it is important to evaluate this correctly since a great deal of play behaviour shows similar postures.
- Hisses are used to avoid frank aggression.
- Growls are agonistic, used both offensively and defensively.

ANNIE

SPOOKY AND BLN

BAKEWELL

LUCY

BINK

ZULU

HOPPY

PUDDY

- Chirrs/chirrups are sounds used when greeting social individuals.

Sociable behaviour
- Sleeping together
- Grooming each other
- Rubbing against each other to exchange scent (the animal that is able to elicit the first rub is allegedly the more dominant)
- Friendly greeting after a prolonged absence
- Play behaviour

Has that helped at all or are you even more confused? You certainly will be when you try to unravel the complex nature of status and hierarchy in cat households. There appears to be a real drive amongst researchers to prove and understand 'rank' in cat groups. Let's face it, they are not small dogs, and trying to turn them into a species with a pack structure is doomed to failure. Whatever cats try to do when they are living in a group is merely an indication of their ability to adapt to an unnatural way of life. If you read all the textbooks you will see that linear hierarchies have been observed in multi-cat households, but again this is a massive oversimplification. There is also some research evidence to indicate that linear hierarchies may shift depending on outcomes of agonistic encounters. There is no question that the more anxious an individual the less likely it is to be dominant. The stability of a group will always depend on the motivational levels for feeding, access to owners and other activities. For example, cats fed periodically can tend to be more aggressive than cats fed ad lib with a plentiful supply of food.

Aggression itself doesn't dictate the 'rank' so much as the

response of the challenged cat. Watch one of your cats enter a room then start to watch the others present to get some grasp of his status within the group. The highest-ranking cat (often the oldest and the heaviest) tends to spend more time in solitary behaviour such as resting, eating, climbing, grooming and scent marking. I shall never forget a household containing seven cats where their every movement was dictated by an ancient cat with kidney failure who spent his time living and sleeping under a desk in the study. There was absolutely no direct intervention but his influence was extraordinarily powerful. It is also worth remembering that inter-cat problems become relevant at the onset of social maturity in the individuals within the group, which can occur between eighteen months and four years of age. This will explain why many owners report that their cats' relationship has changed dramatically at or around this time.

Trying to establish status and rank in any group of cats is a complicated task and, frankly, there are many occasions when I will study a group and fail to see a concrete hierarchy at all. It is probably easier to accept each group as different and certainly, in my job, never to presume a structured hierarchy, linear or otherwise.

Tinker and Sinbad – sibling rivalry

We always try to do the very best for our cats, even when we are making the initial decision about pet ownership. The general advice given regarding kittens tends to be that they should be purchased or acquired in pairs. This is particularly relevant for those of us out at work during the day, enabling two adorable little kittens to be company for each other. Sure enough, for the first year or so they are perfect companions and they play together, groom each other and sleep all curled up in one big fluffy ball. If you are very lucky this relationship will merely cool as they get older and develop into a mutual understanding. Occasionally they will take pleasure from each other's company but mostly they will agree to disagree. Pat and Jeremy were very *un*lucky.

Pat and Jeremy lived in a quiet cul-de-sac in a sleepy town in Hertfordshire. They had two delightful daughters, both under five years old, and a lovely home that mirrored their devotion to their children. The household appeared to be the perfect picture of domestic harmony and family life, apart from the tin foil, the cardboard, the pepper, the orange peel . . .

Pat had called me on the advice of her veterinary surgeon. She had happened to mention to him, out of desperation, that she had a persistent problem with both her twelve-year-old cats soiling in the house. She had always assumed that she was powerless to prevent it and the whole situation had almost become a way of life. The family, however, were starting to get seriously fed up with the constant upheaval. She was very keen, if somewhat sceptical, when her vet said that it was entirely likely I could help. I arranged to visit her to discuss the problem further.

I instantly took to Pat and Jeremy and their beautiful children. I also adored their lovely tabby cat brothers, Tinker and Sinbad. They were twelve years old when I met them and they were very different characters. Tinker was clingy and vocal and persistently approached Pat for attention. Sinbad was quiet and rather independent, choosing to spend time on his own rather than interact with the family. Both cats had been a little shocked by the introduction of children to the household and adopted a tactic of avoidance at all times. It was great chatting to Pat about Tinker and Sinbad. It was obvious that both she and Jeremy loved the cats deeply and considered them to be incredibly important members of the family. They had lived in three different homes over the years and I saw photographs of the brothers in various poses as kittens and youngsters and mature adults.

It was probably this intense love for their pets that allowed Pat and Jeremy to be so tolerant of their problem. It was apparent from our discussion that Tinker was squatting and passing urine all over the house on a regular basis. 'All over the house' meant literally that. There wasn't a room in the place (or any previous property) that hadn't been anointed with his pee. To add insult to injury, Sinbad had been spraying urine just about every day too. I asked how long this had been going on and my heart sank when I was told 'about ten years'. I was in the presence of a pair of cats that had spent their entire adult life in a distressed state. I mentally rolled my sleeves up and prepared to find out why.

There was certainly a common denominator in all three homes and that was a high cat population in the vicinity. Both Tinker and Sinbad found the presence of other adult cats upsetting and this was a potential cause of their behaviour. After some time I established (with the help of the referring vet) that Tinker had a chronic stress-related urinary tract problem. This had been rumbling on for years and it needed to be addressed with lifestyle changes as well as medication. Sinbad's urine spraying related to a sense of insecurity in the home so we looked firstly at the implications of the cat flap. Were invading cats gaining access? Our initial plan included blocking up the cat flap, dietary changes, increased stimulation and various other labour intensive schemes that would provide a stronger perception of safety indoors. All my clients keep in touch for an eight-week period after the initial consultation so that I can supervise their progress and make changes if things don't go according to plan. One telephone call early on from Pat confirmed my worst suspicion – both cats got worse when Jeremy blocked up the cat flap. This could only mean one thing. For Tinker and Sinbad, the enemy was within the household all the time. I had had a strong feeling that this was the problem but had been met during the consultation with a firm and emphatic 'no' when I suggested separating the two brothers. It was absolutely out of the question. Sinbad was Jeremy's cat and he wouldn't dream of parting with him. Tinker was Pat's cat and she felt equally responsible for him. Any suggestion of rehoming (or worse) would merely represent neglect and betrayal.

I didn't want to lose their trust or friendship so I continued to try every technique I could think of to make the relationship work. I was becoming deeply concerned because I knew I was not doing the right thing for the cats but merely pursuing the unattainable and prolonging their agony. No matter how hard I tried I could not convince Pat and Jeremy to let go. These cases are some of the most difficult I have to deal with. How on earth do you tell caring owners that the greatest act of love they can bestow on their cats is to say goodbye?

Over the next few weeks a genuine friendship developed between me and Pat and Jeremy. By now it was obvious to them that I was putting everything into the challenge and they were starting to see patterns of behaviour that I had predicted. Suddenly they were beginning to realize that the two brothers were as incompatible as I had suggested. I asked them both to grant me one favour. It was agreed that Tinker would spend a short period of time in the vet's own comfy cattery. During that time routine urine samples could be taken and his behaviour would be observed by the trained staff. Sinbad would remain at home and Pat and Jeremy would watch him to see if there was any noticeable difference in his behaviour. What happened next was pretty convincing even for them. The vet reported that Tinker was passing free-flowing urine with no blood and he was using his litter tray. He appeared relaxed and used to play with the nurses in the cattery. Sinbad was also transformed. He did not spray urine once in his brother's absence and he became friendly and attentive, even allowing the two children to stroke him. Unprecedented!!

As Sinbad lay curled up asleep on Pat and Jeremy's bed (another first since he had been a young cat), his owners lay awake with a heap of conflicting emotions. They felt guilty yet relieved; Sinbad was a lovely cat to live with and there was no urine anywhere in the house. They could enjoy his company rather than follow him around with a bottle of surgical spirit and a piece of kitchen paper. What made them feel even guiltier was the fact that they didn't miss Tinker at all. A number of very tearful telephone conversations followed. Both Pat and Jeremy had finally accepted that Tinker and Sinbad were making each other ill. We now had a very difficult decision to make: try to find a home for a twelve-year-old cat with a history of inappropriate urination or put him to sleep. I shall remember that week as a very difficult time; I felt their pain. Just as Tinker's days appeared numbered he got a reprieve. Pat's elderly aunt heard of the dilemma and offered to take Tinker. Just three days later he was taken to his new home and after a short period of adjustment he settled well. He was now provided with the single-cat environment he had always wanted and he could finally relax. His symptoms disappeared and he used his indoor litter tray consistently. Sinbad continued to flourish and Pat and Jeremy took every opportunity to visit Tinker in his new home. At last, everyone was happy and an important lesson was learnt. Sometimes you have to love your cat enough to let him go.

Stress and the multi-cat household

As I have a multi-cat household myself, it must appear strange that I am so anti the whole concept. I am bound to have a jaundiced view about multi-cat households because I spend my life picking up the pieces when they go wrong. Over-grooming, excessive vocalization, idiopathic cystitis and bulimia, for example, are all potential problems resulting from chronic stress. Chronic stress also compromises the immune system and can result in the development of disease. A pathological state of anxiety or depression can also occur in individuals with the inability to exhibit adaptive coping strategies for constant antagonism. I am unusually sensitive to tensions between cats and tend to see problems in all multi-cat households even if they appear to be ticking along relatively peacefully. At the end of the day no life is ever perfect and a degree of antagonism between cats is healthy, normal and perfectly acceptable. It is all a matter of degree.

Hints for a harmonious multi-cat household
Many cats take great pleasure in social interaction with their own species and it's always worth remembering that no two cats (or groups of cats) are the same. However, armed with all this new information about cats and their tenuous grasp of the concept of sharing, it is probably worth looking at measures

that can be taken to limit the chances of a dysfunctional cat household.

- Keep an appropriate number of cats for your environment. For example, if you have a two-bedroom house, then five cats is probably asking for trouble. There is no 'number of cats per square metre of floor space' rule but common sense should prevail.
- Choose compatible individuals such as litter mates, probably brother and sister. Two equal age males may dispute the hierarchy when they mature socially.
- Avoid extremes, for example of nervousness, confidence or activity. Such cats may potentially be difficult to live with or find others difficult to live with.
- Try not to keep adding to the stable social group. Every household has a 'one cat too many' number and you may be pushing your luck.
- When you do add to your group, choose individuals that show a history of being sociable with other cats. Avoid those that have been given up for adoption because of indoor soiling/spraying/anxiety-related problems.
- Avoid friendly entire male strays that appear really gentle and irresistible until they have their paws firmly under the table. They do not do well when living with other cats!
- Avoid too many Burmese in the same household. They tend to be extremely territorial.
- Three-dimensionalize your home. Provide plenty of high resting places to enable any individuals to observe activity from a safe area.
- Provide private places. Every cat, no matter how sociable, needs 'time out' to enjoy moments of solitude. Wardrobes

and cupboards are ideal and there should be plenty of choice to enable all members of the group to have their own favourite place.

- Avoid too many cats in your household in densely cat-populated areas. It can create a stressful sense of over-crowding that can easily cause problems.
- Provide dry food for 'grazing' throughout the day or divide it into several smaller meals to avoid any sense of competition there may be if food is only available at certain times.
- Ensure there are plenty of scratching posts to protect your furniture.
- Beds in warm places are worth defending so provide enough for everyone.

I can safely say that (more by accident than design) I have always followed these rules and, apart from the odd jet of urine here and there, I have maintained relative harmony. Multi-cat households can be hotbeds of discord but it's important to get everything into perspective. Most cats agree to disagree and never show any major signs that the stress of cohabiting is getting them down. If you are experiencing a behavioural problem within your multi-cat household it is important to rule out simple causes before getting down to the complicated stuff.

Flora – a simple case

Flora and her mate, Fauna (naturally), lived with Bridget and her partner Larry in a stylish house in Surrey. Bridget was a high-flying executive and she had researched her problem well long before she called me for my help. Flora was wetting on the floor in the kitchen and Bridget was convinced that she had deep-seated psychological problems. When I visited Bridget I was impressed by her array of textbooks and self-help manuals. This lady had read up on the subject and was prepared to give me her assessment of the problem. Whilst this isn't necessarily the normal way to conduct a consultation I thought I would sit back and let her tell me what was causing the behaviour for a change.

Bridget was convinced that there was something sinister going on between Fauna and Flora. They were brother and sister but she knew that blood wasn't necessarily thicker than water in the cat world. Fauna would repeatedly pounce on Flora and attack her and she would hiss and spit at him. Bridget was convinced that this was distressing Flora. She had also read that cats can guard litter trays and she therefore provided three trays (because she'd read my 'one per cat plus one' formula) in separate corners of the kitchen. Poor Bridget had also panicked when she read about the perils of cat flaps and immediately locked the one in the kitchen, convinced it was the devil's work. She had purchased a 'plug-in' from her vet that emitted synthetic feline facial pheromones to reduce Flora's anxiety. Now she was at her wits' end because everything she had done hadn't made an ounce of difference. Flora still continued to urinate, and occasionally defecate, in two separate locations in the kitchen.

When I am consulting I listen carefully but I also observe. I was watching the two siblings all the time Bridget was talking. Fauna was a little apprehensive of me but he still managed to come into the room and sit beside Flora. She ignored him despite his attempts to get her to play. The two of them were only a year old and still very much like kittens in their behaviour. I eventually said that Bridget could let them outside and the two ran off together. It was obvious to me that Fauna really enjoyed Flora's company. Flora was slightly less keen and found her brother rather irritating when she wasn't in the mood. She certainly wasn't frightened of him. During the course of our discussion several important facts were revealed.

- Bridget had always used a wood pellet based litter material.
- Flora always toileted in the same areas unless Bridget put a litter tray there and then she would find somewhere else.
- Flora preferred to go to the toilet outside and only soiled indoors when the two cats were shut in at night.
- Flora had urinated and defecated inappropriately on and off from a young kitten but the problem was getting worse and was now a daily occurrence.
- There were very few (if any) other cats in the neighbourhood.
- When Flora was observed defecating in the tray she would stretch up in an unusual posture and place at least one leg on the rim of the tray.

Now, with your best powers of deduction, what do you think the problem was? It wasn't a relationship issue; they were still only kittens. It wasn't the cat flap; I don't think Flora or Fauna had seen another cat outside. Flora was toileting on the floor because she hated standing on the hard wood pellet litter! She was trying hard to be clean by choosing just two places to go in the kitchen. They were near the trays but she just couldn't bring herself to go in them. The heavier she got the worse the discomfort of walking on the pellets. The one thing that Bridget hadn't done was change the cat litter. Every now and then a cat will object to this wood-based litter material. It's a great product, lightweight and biodegradable, but it's not exactly sand-like, is it?

So Bridget cleaned her floors, changed her litter material and opened her cat flap. The problem was resolved within twenty-four hours and Bridget was delighted that her two babies were not arch-enemies after all. Never forget to explore the simple solution first!

✻   ✻   ✻

My lovely little 'biker-chick' tabby cat Lucy is still very much part of our remaining group of four cats. Sadly she is now unwell. One of the great benefits of my multi-cat household was a collective resistance to outsiders trying to encroach on the territory. With the demise of Zulu, our little warrior, we started to experience a minor invasion. Lucy never really ventured far from the house so it was a great shock when a particularly nasty stray cat bit her. We heard her scream and it was clear that the cat had attacked her in our neighbour's garden. I cannot remember a time when other cats were allowed to get that close. Tragically this chance encounter sealed her fate and she was diagnosed as FIV positive. Feline Immunodeficiency Virus is the cat equivalent of AIDS and cats rarely live longer than a couple of years after diagnosis. She became ill and it was initially thought by my vet that she had diabetes. I was extremely concerned because she had a fluctuating temperature and this is often a sign of a viral problem. When her illness was confirmed I was deeply shocked but also concerned for the rest of the group. FIV can be transmitted via saliva (as it had been to Lucy when she was bitten), but I was pleased to discover that all the others tested negative. Thanks to advice from the wonderful veterinary surgeons at Edinburgh and Bristol Universities, Lucy is receiving a state of the art treatment protocol using Alpha Interferon and she is looking as bonny as ever, well over one year after the original diagnosis. I am quietly confident that she has several good years in her yet.

CHAPTER 6

The Weird Cat

Bink's Story

THE MOST RECENT CAT TO SNEAK INTO OUR HOME CAME IN THE form of a spitting hissing explosion of a kitten. I was working in a veterinary practice in Cornwall at the time and a client had found the kitten in her garden. All attempts to find an owner had been unsuccessful. The client had put her in a box (with the help of stout gardening gloves) and brought her in. The kitten was very wild and frightened and only about five or six weeks old. Everyone felt she was probably semi-feral and unlikely to make a satisfactory pet. Peter has never been daunted by the prospect of stroppy cats and he felt that she was not unlike a female version of Bln. This wasn't that far from the truth because both cats possessed that notorious orange gene that generates both the ginger and the tortoiseshell coat colours. In the veterinary world tortoiseshells are renowned for their spitefulness. Ask any veterinary nurse which cat sends shivers down her or his spine at the prospect of collecting a blood sample from its jugular. Yes, the tortoiseshell every time; lovingly (or not so lovingly) referred to as 'naughty torties'. Gingers and red tabbies also have a reputation for being feisty and it's not unreasonable to make comparisons to fiery redheads.

Reluctantly I agreed to let her into my life in the absence of anyone else willing to take her on. I feel guilty admitting that now because, eight years on, she is loved and cherished and very much part of the family. Peter took charge of her upbringing and, sure enough, she grew up into a female version of Bln. She loves Peter and deeply resents anyone else. Bink has always been the least user-friendly of all our cats. She is a nightmare to medicate and vaccinate and has only recently, since her seventh birthday, allowed me to touch her without looking completely disgusted. Trust me, that's progress!

Bink and I really do talk a different language and occasionally I see her out of the corner of my eye watching from the top of a kitchen cupboard. It's very unnerving to feel like an intruder in your own home. If I try to appeal to her (in my best behaviour therapist way) I merely succeed in appalling her further. Maybe there are just some cats I can't win over. How ironic that it should be one of my own.

It was easy for me to decide what aspect of cat behaviour should be discussed in Bink's chapter. In my work over the years there have been occasions when cases have baffled and amazed me. Many unusual aspects of cat behaviour are well documented, but others remain a mystery. You will find references to them in the behaviour textbooks and journals but it is quite apparent that they are not fully understood. Here are a few examples of some of the weird and wonderful cases I have seen over the years.

Zebedee – the wool-eating Siamese

The weird behaviour that is relatively common in the Oriental breeds was mentioned briefly in Ying and Yang's story. It relates to a type of bizarre appetite that can cause major problems for cat and owner. Humans can have unusual tastes in food, particularly pregnant women, but it is hard for us to appreciate the delicate flavour and texture of a household rubber glove, for example! 'Pica' is a term used to describe the consumption of non-nutritious materials and it describes a habit indulged in by a very small percentage of the domestic cat population. The substances chosen are many and varied but the most delicious (apparently) include fabric, cardboard, rubber, carpet and plastic. Some cats don't quite manage the consumption bit and are entirely happy to lick or chew; polythene and photographs are well up there in the top ten favourite flavours.

Cats today are offered a mind-boggling array of delicious tinned and dry foods with a great number of owners providing further gourmet treats from their own table. Why would a cat find a cotton tea towel more attractive than a bowl of salmon and trout in lobster jelly? The answer to this question is best illustrated by looking at the case of Zebedee the Siamese.

Zebedee was a two-year-old Lilac Point Siamese who lived with his owners, Laura and Dominic. Laura called me one day at the end of her tether, having tolerated her problem for too long. I went along to see her and met Zebedee. He was a lovely cat, very friendly, but when he grabbed my trousers and started chewing I realized all was not well. This was the problem: Zebedee couldn't stop eating wool. He couldn't stop eating cotton either, or linen or leather. His whole life revolved around hunting for and consuming vast quantities of socks, gloves, underwear, wallets, scarves, tea towels, toys – basically anything he could find made of the appropriate material.

Poor Laura had adapted her entire lifestyle around Zebedee's habit. As I entered the door I followed a ritual of removing my jacket and placing my leather briefcase in a cupboard before I could enter the rest of the house. Obviously she had to stop short of asking me to remove my trousers but I could tell she would have liked to since nothing was sacred in Zebedee's eyes. Laura had managed, with military precision, to remove all potentially appealing items from his reach; everything was locked away in cupboards. Unfortunately she had been too efficient and this had led to a new and sinister twist to Zebedee's obsession. He had turned to crime and become the ultimate cat burglar, systematically visiting all the neighbours' properties, gaining access via open doors, windows or cat flaps and quietly removing their possessions. When Zebedee started stealing wallets and mobile phones, causing the local police to get involved (the neighbours were convinced they were the victims of a crime wave), Laura realized that enough was enough.

This was an extreme case of problem behaviour relatively common in Siamese and other derivative breeds. The most frequently consumed material is wool but the habit often generalizes to other natural fabrics. Behaviourists believe there is a genetic element to the behaviour and that, in some way, the susceptible cat's brain works slightly differently from others.

One theory suggests that the act of chewing causes 'happy chemicals' to be released in the brain, giving the recipient a feeling of intense pleasure. That this then becomes addictive is hardly surprising. Cats are fundamentally predatory creatures and normal cat food holds no particular challenge to catch, kill and consume. Zebedee's behaviour exhibited several elements of the predatory sequence; they were just performed a little differently.

The motivation for this behaviour can vary. In Zebedee's case it was fuelled by his apparent lack of alternative activity. He wasn't suffering from an insatiable appetite that needed satisfying. There just wasn't anything else to do that was equally rewarding. He was a highly intelligent cat and this had become a challenging and active pastime with a very enjoyable reward at the end. How could I come up with anything that could compete with that?

The first thing I did was speak to Zebedee's vet. I had a plan but it needed a little help and the vet prescribed medication that worked in certain areas of the brain to 'tone down' Zebedee's desire to consume material. Laura then devoted her time to providing a more stimulating alternative day for Zebedee. We built activity areas indoors for him to explore, with tasty cat treats hidden away for him to find. Laura stopped feeding him from a bowl and secreted dry food around the house in challenging locations. She offered cooked knuckle bones with residual meat on them to give him something more acceptable to chew on. She played games with him, she taught him how to retrieve and she built assault courses for him in the garden. All her neighbours were armed with water pistols and instructed to 'shoot on sight' if he entered their houses. (Zebedee had a large clanging bell fitted to his collar and they could certainly hear him coming.)

Zebedee was a little confused at first but soon rose to the challenge of his new environment. After twelve weeks he was gradually weaned off the medication and continued to make excellent progress. Despite the occasional missing sock, his pica is well under control. Thanks to the hard work of Laura and all her neighbours his life of crime is a thing of the past.

Simon and the DM boot

Not all wool eaters get a taste for other materials but another Siamese I remember with some admiration was Simon. He was another young adult cat who had been eating his way through the household goods since he was a kitten. His owners had adapted, just like Laura, by hiding everything away. Simon remained undeterred; he just spent a little more time exploring his domain and pulling socks out of laundry baskets, opening drawers and cupboards and generally 'going underground' with his habit. Just before his owner contacted me she had noticed a change in Simon's preferences similar to Zebedee's when she found a half-eaten leather belt. She immediately feared the inevitable and instructed the entire household to place all shoes and bags under lock and key. Unfortunately teenage boys are not always terribly accurate about following instructions and 'kicking the DMs under the bed' was not a sufficient deterrent for Simon. For a couple of weeks peace reigned throughout the house. Simon was quiet . . . a little too quiet. Only then did they discover that Simon had experienced a glorious fortnight under the son's bed eating three-quarters of his left calf-length boot. Doc Martens pride themselves on the durability of their product so I had to take my hat off to Simon's tenacity given the challenge of the task.

I worked with Simon and his family many years ago and I would describe him as 'in remission'. Not all of these cases resolve completely. Recidivism occurs in many wool eaters and I occasionally feel powerless to prevent the continued inconvenience for the families. It also has to be said that it is potentially dangerous for the cat, since many of the items consumed can easily cause an intestinal blockage that requires life-saving surgery. Some owners in the past have resorted to providing a small amount of chopped up wool beside their cat's food bowl to satisfy these continuing urges. It's not ideal, but life just isn't perfect, is it?

If you have a cat that licks, chews or consumes something it shouldn't, it is important you seek help as soon as possible. As if potential intestinal blockages were not bad enough, chewing electric cables (another favourite with Burmese and the like) is extremely dangerous and it is not a problem to be taken lightly. Your veterinary surgeon can help you in the first instance and if necessary refer you to someone like myself.

Lily – the self-mutilating Persian

One of the very first cases I ever treated on my own was a petite cream Persian. Her name was Lily and she was the most loving and affectionate little creature you could ever imagine. Her owner had brought her into the veterinary practice where I was working so that a terrible lesion that Lily had developed on her side could be investigated. It looked like a large area of wet eczema and the poor little cat was really bothering it. As far as I was concerned it was a routine consultation for the senior partner in the practice until he called me into his room to 'have a chat'. It seemed that Lily had done this to herself.

Over a period of a couple of months she had systematically licked and licked her right side until all the fur had been removed. She then continued with the constant grooming and eventually broke through the skin and caused this terrible damage. It was agreed that the partner would carry out some routine dermatological tests to rule out a physical problem that might account for the intense itch in that area. In the meantime he asked me to get to know Lily and her owner and see if I could identify a possible stress trigger.

I was incredibly excited about embarking on such a complex and dramatic case. I was also worried that it might prove too much for my limited experience. The results of the tests showed no evidence of an allergy or other dermatological problem so I was asked to work alongside the vet to see what could be achieved with a behaviour consultation.

Lily's owner, Enid, was a middle-aged lady living on her own in a beautiful townhouse in Truro. She had purchased Lily from a local breeder three years previously and had lavished constant love and attention on her. She was kept exclusively as an indoor cat and the relationship had rumbled on with each providing constant company for the other. Then Enid had become involved in voluntary work for a local charity and had found herself spending longer periods away from the house. She had worried about how lonely Lily had become (she was always so pleased to see her when she returned home) and decided to adopt another cat, an older male that had been kept exclusively indoors in his previous home, to keep her company. Despite her best intentions Enid had introduced a real problem into the house. Lily objected strongly to Arthur, the interloper, and appeared extremely fearful of this rather strapping Cornish moggy. Enid was not unduly worried since she hadn't observed any actual fighting. She just felt there was a little jealousy there, particularly as Arthur was such a 'cuddle bunny'. Over a period of time Enid had started to notice a change in Lily. She had become withdrawn and tended to spend time on her own in private corners of the house. A bald patch appeared on her side and the rest is history.

The case seemed relatively straightforward. Lily was an indoor cat with very little going on in her life and her stimulation and affection came from Enid. Providing a cat pal to fill the void left by a human rarely works and Lily had suddenly found that she had not only lost her human companion but gained a feline intruder. Her world had collapsed inwards and her confused brain had tried hard to develop a strategy to cope with her newfound insecurities. Cats often turn to normal behaviour such as grooming to provide predictable and safe activity in times of strain. Somehow Lily had become hooked on the action of grooming in that one particular spot and even the pain that she induced started to become part of the awful addictive process.

When I had visited Enid I had done so without poor Lily. She was unable to return home because of the extreme nature of the trauma to her side and she was being treated with anti-inflammatories and antibiotics to deal with the infection that had developed in the wound. She was fitted with an Elizabethan collar so that she physically couldn't tamper with the area, and kept in hospital under observation.

At the time, little was known about over-grooming problems. They were occasionally referred to as self-mutilation or psychogenic alopecia and the motivation was considered to be complex and neurochemical in origin. (In reality very few of these problems have a purely behavioural cause.) I sought advice at the time and various suggestions were made regarding medication that might help to break the cycle. If only we could get rid of the dreadful itching and allow the skin to heal we could actively start introducing a behaviour programme that would help. The vet agreed to try Valium (this would be frowned upon today because frankly the results are variable and there are now better alternatives) and Lily became the first patient I had treated on anti-anxiety drugs.

Lily became the practice cat for the following month as she took her tablets and her skin repaired. The nurses and I used to play with her in the surgery, out of hours, and she seemed to love running up and down and chasing pieces of string. We had a secure area outside the kennels and Lily would sit there for hours in the evening just sniffing the night air. I became extremely attached to her and knew what I had to do.

Enid was visiting Lily every day and I was taking the opportunity to show her how much Lily enjoyed being a rough and tumble cat. She was never allowed to climb the furniture at home and Enid was amazed when she watched Lily jumping around the surgery like a little monkey. As her skin was healing nicely I took a deep breath and discussed the unthinkable with Enid. Lily couldn't go back to living with her. Possibly if Arthur found alternative accommodation it might be feasible, but I felt that Lily needed more than that to maintain her emotional health. We couldn't keep her on Valium for ever and I really had a strong instinct that she would love to be living a more natural life. Many owners feel Persians need to be indoors because of the obvious practical considerations: a long silky coat mixes rather badly with garden debris. I would actually disagree; I know many Persians who thoroughly enjoy being outside getting down and dirty. Admittedly they are a nightmare to keep tidy but what price contentment?

When I get on my soapbox I can be quite passionate about certain subjects and, to my amazement, Enid was quickly convinced that this was the answer for the beautiful Lily. A client from the practice, Sarah, had recently lost her cat and found herself without a pet for the first time in many years. She was a single lady who lived in a small cottage in a village on the outskirts of Truro and her environment seemed to be the perfect safe introduction to the great outdoors. Phone calls were made and Lily's prospective new Mum visited her in the hospital and had a chance to meet Enid. Friendships were forged immediately and it was agreed that Lily would go to her new home and be weaned gently off her medication under the supervision of the vet and myself. I cannot explain the enormous sense of pleasure I experienced when I saw Lily in her new garden for the first time. I am absolutely convinced she had a huge smile on her face. Lily went from strength to strength and never returned to her over-grooming. Arthur blossomed and became rather too rotund but perfectly content with Enid. Lily and Sarah became firm friends but Lily tended to enjoy the delights of outdoors and was rarely seen during the long summer days. Not every case since has had such a rosily romantic conclusion but I can honestly say it was Lily who really fuelled my addiction for the job.

'Kitty Prozac'

Ever since Lily and her Valium I have been strongly opposed to drugging cats in this way. Over the years I have realized that there are very few behaviour problems that cannot be resolved without resorting to such potent and toxic medicines. The reason why they are used relatively frequently now is that the alternative is often to remove the cat and rehome it elsewhere and the owner cannot face that decision. Admittedly there are some situations where there is a genuine neurological or neurochemical problem that can be helped by short- or long-term medication. In most situations, however, merely removing the cat from the unsuitable environment is sufficient. This really is a very complex issue and I would always recommend discussing the matter thoroughly with your vet before embarking on any drug therapy for problem behaviour.

Tips about drug therapy

- Medication can only be prescribed by a veterinary surgeon.
- Many of the drugs used to treat anxiety or any other emotional state are not licensed for use in cats. Your vet may ask you to sign a disclaimer.
- Always consider drug therapy as a last resort after exploring all other avenues.
- Don't consider drug therapy on its own to treat a behavioural problem.
- All drug therapy should work alongside a programme of behaviour therapy that has been devised by a pet behaviour counsellor or a veterinary surgeon with a particular interest in behaviour.
- Blood tests should always be taken prior to and, in some cases, during drug treatment to monitor liver function.
- Follow advice concerning dose rates carefully. Never stop drug therapy abruptly without consulting your veterinary surgeon.
- Keep any drugs away from children and vulnerable adults!

All of these considerations may go some way to ensuring our cats do not become as reliant on 'happy drugs' as the rest of the population.

Nip – the penis-sucking cat

I often see cats with interesting names such as a doctor's red Burmese called Petechia (meaning a ruptured blood vessel) or the one-eyed black and white moggy called Cooking Fat (switch the first letters around and you'll see where they were going with that one). I cannot hold my hand on my heart and say Bln and Bakewell for example are entirely sensible names and I love hearing new ones that give me a chuckle. A couple of names that made me smile belonged to a pair of moggy brothers called Nip and Tuck. They were eight months old when I visited them and beautiful silver grey tabbies. Their owner, Louisa, called me and I clearly remember her saying, 'I bet this is a first!' Apparently she was experiencing quite a problem with one of her young cats and she was eager to tell me the story.

Louisa had purchased the kittens from a pet shop at the age of seven weeks and had kept them exclusively indoors. She had planned to let them out in the garden after they were castrated but a problem had developed that was delaying their freedom. Nip and Tuck had been romping around the living room one day and Tuck had obviously been a little too boisterous. On further investigation Louisa had noticed that Nip had sustained a nasty scratch to his prepuce. Since it looked exceedingly painful, Louisa took him to the vet who administered some antibiotics and reassured her that it would heal with time. When Louisa returned home with Nip she noticed that he was paying rather a lot of attention to the damaged area. She didn't feel particularly concerned at the time but, over the next few weeks, the situation became rather out of control. Nip became more and more determined to while away the hours grooming around his penis. Actually that wasn't strictly true. Louisa soon discovered, to her complete horror, that Nip wasn't grooming himself at all. He was spending hours every day happily sucking on his penis! Louisa tried to interrupt him (she felt it couldn't be normal) but he merely became a closet sucker and took to hiding away whilst performing this rather unusual pastime. I arranged to visit her the following week.

Nip's story was not as bizarre as you might think. Many cats retain certain infantile behaviour into adulthood. The desire to suckle for comfort remains in many domestic adult cats. This can develop for a number of reasons but an abrupt or early weaning in combination with a 'needy' personality usually does the trick. My Lucy was a classic example. She loved nothing better than to tread up and down rhythmically on my lap with her front paws whilst sucking madly on my sweater. Some cats will merely knead with their paws and dribble uncontrollably. This action mimics that of a tiny kitten stimulating milk flow from its mother's teat. The drooling is an involuntary response to the anticipation of a feed. Both Nip and Tuck were very young when they were acquired from the pet shop. It wasn't entirely clear how old they were and their previous history was completely unknown. Even when Nip first arrived in his new home Louisa would find him curled up with his brother, sucking on the back of Tuck's neck. The damage to his prepuce drew his attention to this area and led him to groom it repeatedly. When he discovered that his penis was a convenient size and shape to suck (and the sensation itself was not unpleasant) Nip merely continued. Not all cats will develop excessive habits of this kind but they can occur in certain individuals to fill voids in everyday life. Many cats choose sleep to fill that void but Nip had other plans. There really wasn't enough going on in his life to provide a suitable alternative or a distraction from sucking his penis.

We devised a plan (including the fitting of an Elizabethan collar) that would provide Nip with an exciting indoor environment to stimulate even the most distracted kitty. Louisa was extremely keen to try anything since Nip had become, needless to say, rather a party trick for all her friends and she was getting pretty fed up with the constant sloppy sucking noises too. Nip was obviously still highly motivated to have a good suck and he tried everything to remove his collar. But Louisa was determined and she took time off work to entertain her kittens. After all, if we got this sorted then Nip could be allowed into the garden to discover the ever changing delights of the great outdoors. That's usually enough to keep most young cats out of mischief.

Time passed and after a very long four months Nip was able to spend supervised time without his collar. It was a long process but Louisa remained steadfast and eventually his penis sucking days were definitely behind him. Not all 'obsessive behaviour' of this kind resolves so beautifully. We were fortunate to be able to provide Nip with plenty of alternative activities whilst consistently preventing the performance of the undesirable behaviour and this is always the key to a success-ful outcome. I would always recommend that owners seek help from their vet and a pet behaviour counsellor as soon as possible if they have a cat that is displaying abnormal behaviour of this kind.

Baldrick – the masturbating cat

Another behaviour that is often adopted as a coping strategy for life's problems is masturbation. This is a taboo area for many when discussing their cat's little foibles and I often wonder how prevalent it really is. Most owners turn a blind eye to Tigger's love affair with the dressing gown but occasionally the behaviour gets so intense it is hard to ignore. If it does become such a persistent habit then it is always a sign that all is not right with Tigger's world.

Baldrick belonged to a young nurse called Chrissie and they lived together in a small cottage on a very busy street. Chrissie worked long hours and Baldrick was always there to greet her when she came home after a tiring shift. Baldrick was about three when I met him and he was a boisterous and affectionate black and white long-haired cat. Throughout the consultation he was knocking things over and jumping up and down like a spoilt child (great fun to watch) and it was hard to concentrate on the tale that Chrissie was telling about his dodgy habits.

Baldrick was kept exclusively indoors because the road outside was just too busy and Chrissie didn't feel comfortable about allowing him out. She provided him with good food and plenty of love when she came home. He had several toys but they all ended up pushed under chairs and sofas in huge piles awaiting the arrival of new ones to replace them. Chrissie obviously loved him very much but there was one part of their relationship that she just couldn't tolerate any more. Every night, without fail, Chrissie would wake to find Baldrick clamped tight around one of her arms or legs with a glazed expression on his face. His pelvis would be thrusting madly and any attempts to remove him prior to completion would result in a burst of frustrated aggression. The end would come and Baldrick would remove himself and disappear to the corner of the bedroom and quietly lick himself. It was not nearly so much fun for Chrissie. She had tried to shut him out of the bedroom but this had resulted in a great deal of damage to the carpet outside as he tried frantically to get to his owner. Chrissie desperately needed her sleep and the nightly experience was thoroughly unpleasant. It was really spoiling her relationship with her beloved Baldrick. Again, great fun for her friends and family who thought it was a hoot, but when you are actually the victim of such behaviour it becomes very wearing.

It was important that we established Baldrick's motivation for the behaviour. Despite the fact that cats are neutered they are still capable of discovering the rewards of masturbation. It can occasionally be triggered by smell so I was keen to establish that Chrissie hadn't used any medication or soaps, perfumes and deodorants that might have added a new pheromone-like element to her scent. I drew a blank and it was soon apparent that we had to look once again at lifestyle to identify the root cause. Baldrick was bored and frustrated. There was nothing to do indoors and the arrival of Chrissie was about as exciting as it got. She would often have to sleep shortly after returning home and Baldrick would lie in wait pending the sight of a naked arm or leg falling out from under the duvet. Then he would pounce and the entertainment would begin.

When you deprive a cat of the opportunity to perform a behaviour that it finds rewarding, it is important to fill the void with an alternative, equally exciting pastime. Baldrick wasn't doing enough cat stuff and he was suffering from 'idle paws' like so many other indoor cats. We had to provide him with entertainment outside his relationship with Chrissie's arm and encourage more acceptable behaviour. Chrissie went to work and built stimulating climbing frames and shelving within her small cottage to give her home a more three-dimensional feel. Cardboard boxes and paper bags were used to good effect and exciting opportunities to obtain food and water were soon introduced. Play sessions became an everyday part of Baldrick's life with emphasis on toys that required the minimum of effort for Chrissie at the end of a busy day (it's amazing how much fun can be had with a couple of feathers tied to a string on the end of a bamboo cane).

Once the new activities were introduced it was time to turn our attention to the night-time naughties. We had to devise a cunning plan that made the behaviour less rewarding or prevented its being triggered in the first place. We established that the masturbation started when Baldrick was given a glimpse of a leg or arm falling out of the duvet. Chrissie was a bit of a star-shaped sleeper and her single bed and duvet meant that all of her was rarely in the bed at the same time. The simple but important addition of an oversized duvet immediately meant that her limbs were less likely to be on view. Unfortunately, restless nocturnal activity means that the duvet doesn't always cover all so we still had to make those arms and legs less attractive. During the course of a consultation you find out lots about the individual cat's likes and dislikes. I remembered Chrissie saying that Baldrick had an absolute aversion to Tesco carrier bags and would jump in the air if one was rustled in front of him. Could we use these bags to devious effect? Sure enough that night Chrissie went to bed with Tesco carrier bags secured round her arms and legs with sticky tape. What a trouper! Several hours later she was woken by an extremely agitated Baldrick who was patting her arm cautiously with his paw. He had a look of absolute horror on his face; this obviously wasn't what he expected at all! After several nights of carrier bag armour and several days of increased stimulation and activity Baldrick stopped even trying to get to Chrissie at night.

I did get a call nearly two years later to say that Baldrick had started masturbating again. This time he had chosen her lodger's arm and everyone was not best pleased. Luckily we were able to dust off the old therapy programme, make a few adjustments here and there, and restore Baldrick's acceptable behaviour once more.

Ceefor – the tail-chasing cat

Tail chasing is rarely the domain of the cat. Border collies and excitable dogs I can understand but I never used to think that it was really a 'cat thing'. That is until I met Ceefor. Her owner called me on her vet's recommendation as all other avenues had been exhausted. Ceefor had started to chase and bite her tail so frequently and so intensely that she needed a permanent Elizabethan collar (what would we do without those things?) and constant supervision. The vet had investigated a medical reason for the problem and had come up with nothing to explain the intense hatred Ceefor appeared to have for her tail. It sounded fascinating so I had a look that very same evening.

Ceefor was a three-year-old tortoiseshell moggy who had been adopted by the family a couple of years previously. They had found her in a local rescue centre, where she had arrived as a stray. She had started to chase and bite the tip of her tail about three months prior to my visit. Her owners, Jillian and Tom, thought it was rather funny initially but soon got worried and took her to the vet. He found that her coat was full of fleas (not always that easy to detect at home) and she had loads of tiny scabs on her tail. He diagnosed a flea allergy and pre-scribed anti-inflammatory drugs and topical treatment for the little parasites. This didn't appear to have any effect at all and

164

Ceefor's behaviour became more and more weird. Ceefor demonstrated beautifully on cue whilst Jillian was telling me about it. Her pupils dilated and her tail shivered violently at the end. She turned to look at it and as she did so it began thrashing wildly from side to side. Ceefor started growling and hissing and then chased round in circles and tumbled head over heels. She grabbed madly at her tail and bit it on several occasions before I distracted her with my feather on a stick and the behaviour stopped. She had always been very easily distracted, either with the good old trusty Elizabethan collar or a game or a gentle touch.

In a consultation it's not enough just to view the abnormal or inappropriate behaviour and then try to find ways to stop it. I have to find out about the cat. I need to know its habits, its relationship with its owners, how it views other cats and how it interacts with its environment. I need to know about lifestyle and I definitely need to know about any changes that have taken place or circumstances that have coincided with the onset of the undesirable behaviour. In this case there was absolutely nothing. I delved and enquired and we all drew a blank. Everything was the same as it had always been as far as her owners were aware. Whatever changes may have taken place were to remain a mystery.

There are always going to be cases that are not run of the mill. There are also many cases where the distinction between a behavioural and a neurological problem becomes distinctly blurred. This was one of them, so I arranged immediately for a referral to a veterinary neurologist. Ceefor went to see him, accompanied by some rather interesting video footage. The results were extremely interesting. Ceefor's skin had become hypersensitive in a localized area due to her immune system's response to all those fleas. She had experienced this strange sensation in her tail and ended up confused, trying to attack whatever was attacking her. Despite her vet's insistence on good flea control Jillian and Tom hadn't quite appreciated how these little creatures can take over your home. A single flea is capable of laying an awful lot of eggs that end up nestling into our luxurious carpets and hatching out in response to the movement around that promises a good meal. Jillian and Tom's house was full of fleas and any treatment for Ceefor was struggling to be effective against a massive onslaught. There was no alternative but to have the whole house treated whilst Ceefor was in hospital to remove the unwelcome visitors.

Ceefor returned (flea free) and a therapy programme of activity and distraction was put in place in the home. With the help of medication prescribed by the vet Ceefor returned to her normal self, but some cases only resolve with a partial amputation of the tail, and I have known one cat who had three partial amputations carried out only to chase her own bottom instead when the whole tail had disappeared. Now that *was* a difficult case!

❋　❋　❋

As you can see, there are some very odd problems out there, not to mention the cat that attacked dogs, the one that moon-walked, the one that ate photographs, the one that pawed constantly at her face or the one that used to eat her owner's underwear! Despite the fact that they all had perfectly rational medical or behavioural reasons for behaving that way, they were still pretty weird! My own weird cat, Bink, is the baby of the remaining group of four and perfectly content with her idyllic lifestyle in Cornwall. Since Bln died she has started to adopt even more of his little habits and I have this feeling that she and Peter will pal up over the next few years. Bink is definitely in line to become Bln II – the sequel. I am pleased to report that she let me stroke her recently without pulling that miserable face so I think I'm wearing her down. By the time she is fourteen I may be able to pick her up!

CHAPTER 7

The Human/Cat Bond

Zulu's Story

SHORTLY AFTER THE DEATH OF HOPPY, THE BIGGEST CHARACTER in our feline family, we started to get a visitor to our cottage in Cornwall. Zulu (so named apparently because he was the feline version of a black warrior) was an intact tomcat who lived with a lady in the village. We had seen him around but Hoppy had put paid to any ideas he might have had about getting near the house. Hoppy would merely fix him with a steely glare, despite the fact that his eyesight was poor by then. Zulu would pretend not to see him and nonchalantly change direction as if he had suddenly remembered something urgent he should be doing back home.

Within days of Hoppy's death I came into the cottage and thought I saw Bakewell sitting on the sofa. I wondered why he looked a bit odd until he turned to face me and I saw we had an interloper. Zulu had arrived and was here to stay and there was nothing anyone could do about it. We told his owner and she said she didn't want him and would gladly hand him over, so he was castrated and became number six cat in the Halls' household. The only two cats that hated Zulu initially were Bln and Puddy. Bln became top cat after Hoppy died but when he saw this muscle-bound black cat he knew his reign would be short-lived. During the next few weeks we had our first experience with indoor urine spraying. Puddy just hated Zulu with a passion but, despite all the antagonism, he remained totally committed to staying permanently. There were always a few arguments, since Zulu really lacked a lot of social graces, but he never lost that look of sheer delight on his face at finding such a nice new home.

I often think about Zulu's previous owner and her apparent lack of bonding with this delightful creature. She was a gentle lady and a true animal lover but she willingly gave up her pet to another home. I have spoken to countless cat owners over the years and the concept of parting with their pets is always a difficult one. There are, however, always circumstances or types of relationship that allow such splits to take place without much emotional upheaval. I have now come to accept that the bond between human and cat varies enormously, and it is this aspect of my work that I find most fascinating.

Most of us fully accept that, to some degree, we are at the beck and call of our feline friends. We know we do stuff for them and spoil them rotten but we do it because it gives us pleasure. We think we live our lives with freedom of choice to decide what we do and when we do it, but do we? Don't you ever wonder who's really in charge in your household? As you carefully prepare Ginger's fifth tiger prawn, do you ever feel that somebody somewhere is taking advantage? I never cease to be amazed by how incredibly clever and manipulative the average pet cat can be. It is rather humbling and intensely amusing to think that we can be so successfully trained by a small furry creature. Cats are supposedly well beneath us on the evolutionary scale but sometimes I wonder.

Just cast your mind back to how it was at the beginning. Within minutes of his arrival the new kitten has endeared himself to everyone with his chocolate-box looks and cute bow-legged jaunts around the house. By the time he is one year old he rejects all but the most expensive gourmet cat food and commandeers the best chair in the living room and the end of your bed. This is still not enough and over the years he learns how easy it is to get you jumping through hoops to satisfy his insatiable desire for pleasure. On his tenth birthday he has the master bed almost entirely to himself since he has developed the knack of sleeping star-shaped whilst expanding to three times his normal size and weight. He eats nothing but freshly cooked chicken breast and can get attention at any time of day or night just by calling.

How could this happen? Why do we allow our every action to be directed by Smokey from the comfort of his radiator hammock? Why are these cats so fussy and so demanding? The simple answer is we make them that way (and we love it). A cat is born with a large number of instinctive behaviours already pre-programmed, but many of the things that our cats do result from their life experiences. In technical terms this is referred to as operant conditioning and it means learning from the effects of an action or set of actions. For example, if you open the door for your cat every time he scratches at it he will learn that this is an effective way to gain entry to the house. Whilst this is incredibly useful, the ability to exercise this talent can be taken to extremes. If your beloved dares to draw breath before eating a meal or walks away after a disinterested look, do you provide alternative goodies that may be more tempting? If that is the case then you will soon have a cat who will only eat hand-fed smoked salmon and drink double cream from a china saucer.

We all have a choice how to deal with this. The option I have personally chosen is to spoil my cats with love and attention whilst still remembering that I have a perfect right to live in my home too. I occasionally reject them if I am busy and if they turn their noses up at the food I offer I merely presume they are not hungry. I am fortunate to have (dare I say) cats that are comparatively well adjusted.

If you feel you are no longer boss in your own home and would like to restore the balance of power, here are a few tips:

- Start as you mean to go on. There are many veterinary formulated foods that will provide your cat with everything he or she needs. The occasional titbit is fine but if your cat's appetite is not voracious and he or she isn't losing weight, don't worry.
- It is perfectly all right to reject your cat's attentions when you are busy elsewhere. He or she will not love you any less.
- If your cat should demand attention at night do your best to ignore it. If you don't, you will end up with dreadful sleep deprivation symptoms and a cat that is very difficult to live with.
- Try to ensure that your cat has other interests apart from you, including visits outdoors whenever possible and activities around the house.

If all the benefits are mutual in the relationship between you and your cat, then it is impossible to see any harm in it. I will admit it is always interesting to meet people who prefer the company of cats to other humans and, given their circumstances, I can often see their point. During my career, however, I have seen many cases where these attachments are not necessarily in the best interests of either party. We all know someone who we think is slightly dotty in the way they behave towards their cat (not counting ourselves, of course) but it is usually harmless fun, like the whole Bln/Peter thing, that gives pleasure to both the cat and the owner. Occasionally these relationships are not quite so healthy and they lead to problems, particularly when either the cat or the human or both become 'over-attached'. It is hard to believe that we can love our cats too much, but when you start exploring the nature of some relationships you realize the pitfalls.

If I were asked to define an over-attachment, I would probably answer: *An emotional bond with a pet that is so intense that it is detrimental to the physical or psychological well-being of either the human or the animal or both.*

These are, as I have said before, definitely 'behind closed doors' relationships that many of us will never fully appreciate or understand. (I only have to think about poor Miss X, whom I described in the Introduction, when I say that.) There are probably thousands of similar 'couples' out there but they don't come to people like me because, usually, they don't see their dependency as a problem.

Veterinary practice staff may see the aftermath of over-attachment problems when the animal dies or is put to sleep and the owner is unable to cope. Some people can exhibit extreme emotion and tend not to progress through the stages of grieving in the usual way that inevitably leads to acceptance of the loss. Only recently I met a lovely lady, Emma, who was still intensely grief-stricken about the loss of her favourite cat some eight months previously. I had the opportunity to speak at some length to Emma, and it was apparent that there had been a great deal of sorrow in her life revolving around a failed relationship and the loss of her husband. Over-attachments can easily occur as the owner turns to a loving and non-judgemental pet to help cope with an unresolved emotional trauma in their life. How well I remember sitting with a bereaved owner trying to comfort her as she told me that the cat had belonged to her dead husband. She then intimated in floods of tears that she herself had assisted in his 'mercy killing' when he became terminally ill. These relationships are some-times not about the cat at all.

During my career as a cat behaviour counsellor I have seen a lot of cases over the years. Most involve indoor soiling, urine spraying in the home and aggression, but only about 10 per cent of all the cases relate to behavioural problems caused by over-attachment. Within that 10 per cent, the most common presentation is unusual cat/owner responses on both sides. For example, a slightly incompetent and nervous cat goes to live with a caring, solicitous, emotional owner and the eventual result can be a 'learned helplessness' in the cat and an over-attached relationship. The owner is so busy reassuring the cat that it will be suitably protected that the poor shivering wreck sees danger everywhere and collapses in a heap – completely unable to do anything without the owner present. The other scenario tends to be a highly intelligent, sensitive cat (for example the delightful but high-maintenance Siamese and Burmese) meeting the same caring, solicitous, emotional owner. The outcome here is very different, as undesirable attention-seeking behaviour often becomes the problem. (We will shortly explore Chester's case, which will illustrate how 'undesirable' this behaviour can be.)

We are now all fidgeting nervously in our seats and tickling our Burmese distractedly under his chin as we ask, 'Is there a particular type of person that ends up in this predicament?' Not surprisingly these cases have some elements in common. Of course all of us will recognize something of ourselves in some of the categories and that is only to be expected. Before rushing out to seek help it is important to remember that problems occur only when many of the elements are present together. The cat also has to have the sort of personality that makes it unable to walk away from an over-attached owner. Here is a list of factors common to all the cases I have seen where problems have arisen as a result of the strength of the bond between owner and cat.

- The owners are exclusively women (worried yet?).
- The owners live alone or with a partner or companion with whom they spend little time.
- The owners have been or are on Prozac or a similar psychotropic drug or have been treated for a psychological problem. A large percentage of them have experienced a bereavement or divorce.
- The owners are anthropomorphic about their cats, referring to them as if human. (Many conduct the consultation by talking to the cat rather than make eye contact with me.)
- They don't go on holiday or visit friends or family overnight because they are reluctant to leave their cat.
- The cats are kept exclusively indoors or only allowed restricted access to outdoors under supervision for reasons of 'safety'. (The owners worry that their cat would be

exposed to unacceptable dangers if he or she were to go outside.)

- Many of them refer to themselves as perfectionists, eager to please and desperate to do the right thing.
- Their lives often revolve around the daily requirements of their cat. If working, these owners make incredible provisions for the well-being of their pet during their absence and they can't wait to return home.

I really don't want to give the impression that I sit in judgement whilst firing questions relating to these criteria. 'Are you, or have you ever been, on Prozac?' This just doesn't happen and frankly if it did I would fully expect to be shown the door. All the information is volunteered to me. It may or may not be relevant but I listen because knowing the intensity of the emotional dependence is vital if I have to tell a person that she needs to change her relationship with her beloved cat. I must admit I still struggle with the whole 'small human in furry jacket' thing. Everything the owner observes the cat doing is interpreted in human terms. This is usually the basis for the whole relationship and I admit I am guilty of it myself in my more fluffy moments. But I know it's not real – it's just fun. However, I am frequently confronted with comments such as, 'Bubbles tells me I'm a silly mummy and he worries about the company I keep.' Only a matter of weeks ago I said, rather ill-advisedly, 'Cats are not small people.' As soon as I said it I felt I should get my coat and leave as the reply came, 'Yes they are!' In these circumstances I must admit I employ a technique that my peers would frown upon. Alienating this client would have been pointless and wouldn't have helped the cat at all. So we hold Alice's hand and go 'Through the Looking Glass' and discuss a programme of therapy based on *her* interpretation of the relationship and the cognitive processes of the little furry human. Without my actually acknowledging or denying the erroneous nature of the owner's feelings the cat gets better and I've done my job.

These over-attachments can be the most challenging of cases because an understanding of the owner is so critical to success. The most essential thing that I have to remember is not to judge. Everyone has different life experiences and I cannot possibly say that, given the same circumstances and personality, I would do any better. Having said that, there are some instances where I wonder. For example, one lady left the heating on all night in case the cat got cold and stayed awake for most of it because it was too hot to sleep. Another lady placed seven bowls of food down every day containing different varieties, in case the cat wasn't in the mood for any particular flavour and would therefore go hungry. (Needless to say that poor woman had been diagnosed with an eating disorder herself.) I also remember one dear old lady who used to get up at 3 a.m. to cook fish because that was when her cat asked for it.

Before falling over in complete incredulity, remember that many people are in desperate need of something to care for and something to love. To an extent we all have some similar, but perhaps not so extreme, elements in our own relationships with other people or pets. I'm not there to judge but to try to help.

Chester – the attention-seeking cat

Now let's talk about Chester and his rather unsociable habit. Although usually seen as a territorial or anxiety-related behaviour, spraying urine is an extremely effective way of getting attention! Chester was a three-year-old neutered male

Siamese and he lived with two other cats and his owners, Rebecca and Matthew. He had a cat flap to allow access to outdoors but this was shut at night and when the owners were out 'for safety reasons'. Chester had a rather unsettling habit of jumping into cars and travelling around the country with his newfound friends. Rebecca had become so distressed at one point, after a particularly lengthy journey to Norwich, that she shut the cat flap for good and confined Chester to barracks. He now had to turn to Rebecca for his entertainment. The other cats in the household were no substitute at all. As far as Chester was concerned human company was now the most important thing in his life.

There was no question he was Rebecca's favourite cat. She felt he had a vulnerability that the others didn't have and he needed her to look after him. Rebecca's partner, Matthew, was away on business a great deal so Chester became a more regular feature in her bed than him. I was given a piece of information during the consultation that I could easily have done without. When Rebecca left the room briefly, Matthew was at pains to point out that I had better sort this problem out PDQ because it was putting an enormous strain on their relationship. Apparently Chester was not prepared to give up his position beside Rebecca in the bed just because Matthew had come home. Chester was proving to be an extremely effective contraceptive as a single strategically placed claw interrupted all foreplay. (Great. Just the sort of pressure I needed to come up with the goods.)

Well, it was apparent that the relationship between Chester and Rebecca was proving a problem for Matthew. However, the element that was an issue for Rebecca was even worse. Chester had really started focusing on her since his enforced confinement indoors. He went everywhere with her, constantly miaowing and weaving in and out of her legs. Rebecca talked back and picked him up and put him down and generally spoilt him rotten. The real unpleasantness started when visitors arrived. Rebecca was a perfect hostess and she was always attentive to the friends and family who came to see her. Chester, finding himself without her full attention, started to experiment with methods to return her focus to its rightful place. Banging pictures on the wall and knocking ornaments off shelves was quite effective, but spraying a jet of urine into her face was an absolute winner! Chester would jump up onto the table if dinner was being served or onto her lap if she was relaxing with friends. He would then elegantly lift his tail and squirt a stream of warm urine onto the object of his affection. It worked every time. Rebecca was distraught. She loved Chester but found this behaviour absolutely disgusting. As is often a feature of such relationships she saw herself as totally responsible for it and felt enormously guilty. What had she done wrong?

Whilst constantly trying to deflect Chester's quivering tail I tried to explain to Rebecca what was happening. The combination of a very loving, over-protective owner and an intelligent, highly sociable and manipulative cat had created an intense relationship. Chester constantly approached Rebecca for all his interaction and stimulation. His requirements for attention had increased since she had shut the cat flap and deprived him of other activities. Rebecca was a totally compliant owner and she responded to his requests every time. The more she gave, the more he wanted. This started to become a problem on those occasions when Rebecca did not focus all her attention on him and Chester became extremely frustrated. He tried his previous approach to get attention. No joy. He tried harder. Still no joy. So he thought laterally and experimented and eventually came up with the ingenious urine-spraying trick. Cats spray urine as a reactional gesture when they are anxious or in situations of conflict. Chester soon learnt that it achieved its goal by getting Rebecca's attention and it also diffused his frustration at the same time. Voilà!

There was an answer to this problem but I really needed to have Rebecca on my side. I needed to show her how to love Chester differently. Not less, I was keen to reassure her, just differently. Rebecca had to try to control her concern for Chester and understand that whilst he continued to be constantly focusing on her, it was just as disruptive and stressful for him as it was for her. Chester needed to be given more freedom. We decided to ask all the neighbours to be extra vigilant about driving away with him. The locals soon got into the habit of doing a 'Chester check' before setting off. Rebecca now had no good reason to confine Chester indoors and his hunting trips were reinstated. Despite his affection for his owner, she was no substitute for the pleasure of sitting staring at a hedge for hours on end waiting for a mouse.

Chester continued to be kept in at night but I suggested that the provision of a sheepskin hammock on the radiator in the spare bedroom could well be a tempting alternative to the marital bed. I never like to deprive an attached cat of access to his owner without offering an alternative. His excursions outside were a big diversion but night-time still posed a problem. In this particular case I felt that it would be better for everyone if Chester were shut out of the bedroom at night. Luckily his ravenous appetite and heat-seeking tendencies eventually made the radiator hammock and a bowl of crunchy biscuits a perfect substitute for a night with Rebecca under the duvet. Obviously, the transformation didn't happen overnight. Many tears were shed and a great deal of patience and determination was needed to ignore Chester's initial attempts to gain access to the bedroom. A burst of frenzied nocturnal activity gradually gave way to a withdrawal to the comforts of the guest bedroom and the delights within. If it hadn't been for Matthew's support at this time Rebecca would never have been strong enough to enforce the change in routine. For some reason he played a very active role in this part of the therapy programme!

There were obviously still occasions throughout the next few weeks when Chester employed his previously infallible technique to obtain attention. Rebecca had very strict instructions to ignore this and not to respond as she had previously done. This sounds great on paper but when you are dripping with cat urine and desperate to go 'Yuk' and wipe it off it is extremely difficult to maintain a nonchalant air. We agreed that Rebecca would wear old clothing for the necessary period and just hang on in there until things improved (whilst keeping her mouth and eyes firmly shut at those dangerous moments). Occasional weakening and gestures towards him would have made it even more difficult to stop the spraying since it would have represented a very tempting regime of intermittent reinforcement – any gambler will tell you how addictive random reward can be! At times when Chester was behaving well and not all over her like a rash he was rewarded with games and affection. Her amazing efforts were eventually rewarded as Chester, after several enthusiastic attempts to gain attention, one day left the room with an audible huff and never sprayed urine on her again. Wouldn't it be great if things always worked out that perfectly?

For over-attached owners, to ignore their pets is the most difficult task they will probably ever perform. There has to be no eye contact, no verbal communication and a body language that is unfamiliar to the cat so that the signal 'not at the moment thank you' is clear and can be understood. I would love a pound coin for every time an owner has misunderstood my instructions and turned to their beloved pet, wagging their finger and saying, 'Now stop that! I've told you I'm ignoring you because your behaviour is absolutely unacceptable.' Programmes to modify or change the undesirable behaviour often fail as such owners, without the wonderful support of a Matthew, find it impossible to reject their beloved pet. And even when they have gritted their teeth and gone through with the therapy, I know that many reformed characters slip back to old ways as soon as my back is turned. Such is life.

Billy – the over-attached cat

Another case I recall illustrates an unusual variation on the theme of over-attachment. This was a very one-sided relationship with all of the intensity coming from the cat and not the owner. It also illustrates once again that not all cases have a happy ending. Billy was a young castrated moggy, grandly referred to in the trade as a domestic shorthair. He was homed by an animal charity at the age of five months to a couple out at work during the day. All went well until, at the age of nine months, he started to defecate on the floor, apparently involuntarily, when he greeted his owners in the evening. The owners returned him to the charity as they found this impossible to live with. He then stayed at a foster home with an extremely knowledgeable and experienced lady with the intention that the habit would be broken and he would then be permanently homed elsewhere.

Billy proved to be a handful. He paced and miaowed continuously when he was not with his foster owner and he defecated when he saw her, even on her feet. The faeces appeared to be voided without conscious control and the act was preceded by vigorous rubbing around his foster owner's legs. When he was close to her he hyperventilated and appeared over-excited. He even used to try to climb up her clothing to place his head inside her mouth! He never stopped actively seeking her company and he never appeared still or relaxed. His attentions were definitely focused on this one person.

I was perplexed. I had no idea how he could have developed what appeared to be such a severe bonding disorder. Unfortunately it just wasn't clear what his motivation was for this behaviour. It may have been frustration and fearful insecurity or a bizarre form of attention-seeking behaviour that had received inadvertent reinforcement from the foster owner. I also couldn't rule out the possibility of a medical problem, maybe excess or insufficiency of a certain chemical in his brain. I desperately tried to research this case with colleagues but everyone was baffled. His vet had prescribed an antidepressant that is often used to treat anxiety-related problems in cats but it only made things worse. Any behaviour therapy I tried was perfectly executed by Billy's foster owner but had no positive effect whatsoever on the situation. After exhaustive research I came up with a plan to feed a particular diet that might enhance the production of the chemical in Billy's brain that I felt could be the root cause. A full blood screen and neurological referral were planned but I was too late. Billy's foster owner had come to the end of her tether. She had no further emotional reserves to deal with this intensely distressing problem and she requested that Billy be put to sleep. I will never know why Billy behaved as he did and I have never seen another case since that is remotely similar.

George – the under-attached cat

Under-attachment can also be a problem for some people but it is a very subjective issue since one man's under-attached cat is another man's normal independent moggy. One case that was actually referred for aggression illustrates this point. Sometimes you have to look beyond the alleged problem to see what's really going on.

George was a two-year-old male neutered cream-coloured British Shorthair. He lived with his owners, Michael and Sandy, and their two teenage children in a large two-storey house. It was a stylish home, all beige and gold, and George co-ordinated beautifully. I usually start my consultations (unless there is a good reason not to) by saying a quick hello to the patient to see their response to strangers. George was sitting, with his paws tucked under him, on a sofa in the far corner of the living room with his back to the family. He was firmly indicating that any interaction was not desired at that time but I felt I had to test the strength of his conviction. As I talked softly to him I could see his ears twist so I crouched down and gently stroked the top of his head. If I had smeared him with some obnoxious gook he couldn't have looked more disgusted. As I stroked further down his back his skin rippled and he looked ready to vomit. Fair enough. He obviously wasn't enamoured of strangers but he didn't object in a demonstratively aggressive way.

I left him alone, much to his delight, and started to listen to Sandy as she discussed the problem and George's lifestyle. George had access to outdoors but he was very reluctant to explore and would only go out if the family were in the garden on a warm day. He had a good appetite and was fed a complete dry cat food and some tinned food twice a day. He was very unsociable with people (really?) and withdrew from any physical contact. However, he loved to play with fur mice and the children used to throw these around the house for him.

The aggression related to a behaviour he had indulged in ever since his arrival at his new home as a twelve-week-old kitten. He loved to attack people's feet, usually after playing with one of his fur mice. When he was tiny it was considered amusing; in fact the family were thrilled that the cat showed some interest in them and actively encouraged it. He was such a stand-offish animal at other times that they relished the inter-action. The pattern of attacks continued and his bites and scratches became painful as he got older and stronger. The children soon lost interest and by the time I was called in they had stopped playing with him completely, hoping that this would prevent further attacks. Unfortunately it hadn't and the family was becoming frustrated. At this point in the consul-tation it was apparent to me that there was a more important underlying issue to all this. Everyone felt that George just wasn't a member of the family. Frankly, they couldn't see the point of him at all. He was snooty and stuck-up most of the time and when he wasn't being aloof he was ripping their feet apart. They had started to resent him and focused on the aggression to illustrate his unsuitability. Basically he had failed to conform to some unwritten Trades Description Act for pets. They almost felt they should go back to the breeder and ask for a refund because he was faulty.

George was certainly very unsociable in character, but he also lacked confidence, which prevented him from filling his day with loads of challenging activities. When he did play a game he got over-excited and hurt people. This had become a learnt behaviour as a result of the family rewarding it so much when he was a kitten, just like Monty in Chapter 3.

Everyone, including George, was miserable and things had to change. The goal was for the family to get pleasure from him and vice versa in a positive but non-harmful way. The attacks had to stop but we had to replace his one true pleasure with something equally rewarding and more socially acceptable.

The first change was George's feeding regime and we created some novel feeding opportunities with his dry food that avoided the obvious chance to attack his owner's feet at meal-times. The majority of his dry food was placed in various locations throughout the conservatory; quite obvious places at first so that he got the hang of it and then in more hidden spots. The teenage children were enlisted to build a cardboard assault course with lots of hiding places for food and catnip. They reassembled his old activity centre scratching post that they had stopped using when George was about six months old and added a high vantage point from which to watch the birds outside on the bird table.

The games he so enjoyed would now take place at the end of a long bamboo cane so that the children's feet were so far away from George that they were no longer a target. George was fascinated by running water, so they installed an indoor fountain in the conservatory, which he loved. Feng shui water features are a great invention for a bored cat!

As he became more and more enthusiastic about life in general and all the new activities, George's confidence increased. After about three weeks, as instructed, his owners left the conservatory door open one day (very casually) and allowed him to make his mind up whether or not he went out. He did, tentatively at first and at his own pace, with no intervention from his owners.

George improved enormously and the aggression disappeared overnight, replaced with fun and games, exploring the garden and challenging his newly found confidence. The family felt more affectionate towards him because he appeared lively and relaxed in their presence. They were instructed not to approach George but to leave him to do his own thing. He was not being subjected to human contact against his will but discovering for himself how rewarding interaction with humans could be. It was just a matter of time before he explored the possibilities of direct tactile contact.

Several months later I had a telephone call from Sandy. She was extremely excited and I had to work hard to figure out who it was making the call. All I could hear was a frantic whisper saying, 'George is sitting next to me! He just jumped up on the sofa and he's sitting right next to me!' Now, this may not mean much to you but for George and Sandy it was a milestone. The relationship had given so few rewards and the mere fact that George had decided to sit next to his owner was the most marvellous result. I whispered back that I was thrilled (after all I didn't want to frighten him off) and said that I hoped it was the start of a beautiful and long-awaited friendship.

Twinkle – a distressing case

There are occasions when I see things in my work that haunt me for a long time. One particular case showed how complicated certain human/cat relationships can become. Whilst there is a degree of human psychology incorporated in the training of a pet behaviour counsellor, we are ultimately not psychiatrists and every now and then a situation arises that emphasizes this point all too well.

I will always remember the telephone call from a veterinary surgeon in Hampshire. He used to refer cases regularly and we had got to know each other reasonably well. He had become particularly worried about a middle-aged cat, Twinkle, who had been suffering from a skin problem for approximately eighteen months. He had been referred to a dermatologist and many tests had been conducted, all failing to achieve a diagnosis. Twinkle apparently persistently scratched at his ears, leaving them sore and bleeding. The dilemma was that there didn't appear to be any physical reason for this irritation. As I mentioned in the last chapter, these self-mutilation or over-grooming problems are predominantly physical rather than psychological. However, I respected this vet enormously and, reading between the lines, he obviously had a gut feeling about this case. His actual words were, 'Do me a favour, Vicky, go and have a look and see what you can find out.' So I did.

Miss Frobisher was delighted to receive my telephone call that same afternoon. We had a brief chat and I too started to get a strange sensation in the pit of my stomach. I have learnt over the years not to ignore this intuitive warning light and I was getting a sense that this might be a relationship issue of some kind. I made an appointment to visit Miss Frobisher but I arranged the consultation for the following week. In the meantime I asked her to prepare a daily diary for me of everything that happened in a typical day for Twinkle. The nature of these diaries can be very revealing and a useful tool to review during the initial consultation.

When I arrived at her small bungalow in a quiet residential cul-de-sac I was surprised to see all the curtains drawn. When I knocked on the door I feared that she had overslept and that she would greet me all bleary-eyed and confused. I was wrong. Miss Frobisher greeted me enthusiastically and escorted me into her small living room. I allowed my eyes to adjust to the gloom and I did a quick appraisal of the surroundings. The furniture was arranged in a very unusual fashion with two chairs in front of a settee, all facing the small television in the corner, almost like the first two rows in the cinema. There was a small table and two chairs against the wall and a litter tray positioned in the corner. In front of the fireplace was a food and water bowl and the room contained very little else apart from a tiny cat wearing a plastic Elizabethan collar around his neck. Twinkle was peering at me from under the table and he immediately came out to greet me. He alternated between chirruping and purring and pacing round and round the two chairs in the centre of the room like a tiny car on an imaginary Indianapolis racing circuit. Miss Frobisher ignored him whilst passing me the typed diary of the previous week for my perusal. She explained that, as Twinkle was fitted with a protective collar to prevent self-harm, she had confined him to this room for his own safety. I started to read the diary, more as a means of distraction than anything. The first day went something like this:

7.16 a.m.   Got up.
7.24 a.m.   Entered living room. Emptied contents of litter tray, consisting of one bowel motion and one urination.
7.35 a.m.   Fed Twinkle and provided morning medication therein.
8.00 a.m.   Had breakfast and left for work.
2.12 p.m.   Returned home. Emptied contents of litter tray, consisting of one urination.
2.30 p.m.   Cleaned Twinkle's collar and removed dried food.

2.45 p.m.   Washed Twinkle's face.

5.30 p.m.   Fed Twinkle and provided evening medication therein.

7.34 p.m.   Emptied litter tray, consisting of one urination.

10.48 p.m. Went to bed.

The diary went on in the same pattern, with Miss Frobisher filling her day with emptying the litter tray, feeding and medicating Twinkle and cleaning his plastic collar. This daily routine disturbed me greatly. I started to talk to her and a devastating fact became apparent. Miss Frobisher had kept Twinkle confined to that single room with a plastic bucket on his head for nearly eighteen months. He had not seen daylight (too distracting for him) and he had not left the room (he might escape through the cat flap). I asked Miss Frobisher if she would be kind enough to make me a cup of tea. I needed to be left alone so that I could measure the room. Twinkle had been confined to an area no larger than twelve foot by ten for nearly one fifth of his life.

When Miss Frobisher returned to the room I took a deep breath and a big chance. I asked her if she trusted me. She had been very enthusiastic and forthcoming about all her efforts to look after her pet and the care she had lavished on him and I had replied encouragingly. She obviously felt that I understood what a dedicated owner she was and she confirmed that she did indeed trust me. I explained that Twinkle's ears had healed and that there was no reason whatsoever to maintain the protective collar or his enforced confinement. (At this point the consultation came to an abrupt halt as Twinkle passed a bowel motion. Miss Frobisher immediately donned rubber gloves to remove the offending object from the tray with the aid of a trowel and a plastic bag. She sealed the bag, left the room, deposited the item in the dustbin by the back door and then returned. The conversation continued as if nothing had happened.) Miss Frobisher appeared most concerned; she asked, 'What would happen if he started scratching again?' I explained that it would be perfectly normal for him to scratch and wash because he hadn't really had much of a chance to clean himself for a very long time. She was not convinced. I even offered to trim his claws so that he wouldn't be able to hurt himself. Reluctantly she eventually agreed, luckily before I had turned nasty and insisted.

I shall never forget Twinkle's little face when I removed the collar. He looked disorientated and frightened. I reassured him and touched him and talked to him and allowed him to adjust to this radically new sensation of increased movement and peripheral vision. He started to wash. He washed and he washed and he washed. It was good to see. After all, there was a lot of catching up to do. As he started to walk rather gingerly around the room I carefully pulled the curtains back. Both Twinkle and I squinted and blinked at each other. My eyes were watering at this point so I probably imagined that Twinkle looked at me and opened his mouth in a silent miaow of thanks. Whilst this was happening Miss Frobisher was sitting on the settee with her knees clasped in her arms looking concerned. I carefully opened the living room door (not before I had blockaded the cat flap, at Miss Frobisher's insistence, with every piece of furniture I could find) and encouraged Twinkle to stretch his legs and explore.

After several hours Twinkle looked relaxed and settled on his owner's bed. Miss Frobisher had also relaxed sufficiently to continue her monologue about her abilities as a cat owner and her most caring nature. She told me about her previous cat (also called Twinkle – what a coincidence) and described how he had been on medication for sixteen years before he died in his sleep. Miss Frobisher had been heartbroken but had consoled herself with the acquisition of Twinkle (the sequel) from a local rescue centre. How tragic it was that he too had become poorly so soon after his arrival in his new home.

I worked with Miss Frobisher and Twinkle over the next eight weeks. Eventually he was allowed to resume his activities outside and he started to live a relatively normal life again. I recommended a strict regime of flea control and visits to the vet to ensure that he would have no cause to start scratching his ears again. I had occasion to speak to the referring vet some years later and asked him if he had seen Miss Frobisher and Twinkle recently. Apparently Twinkle had been trouble free for nearly a year after my visit until he became ill and died of kidney failure shortly afterwards.

I have often thought about Twinkle and Miss Frobisher. I try desperately hard to understand situations that appear to defy explanation. How could she possibly have thought that she was being kind to Twinkle? Was her apparent cruelty the symptom of an undiagnosed psychiatric disorder? There is a recognized condition referred to as Munchausen's syndrome by proxy, where parents cause illness in their children to gain attention for themselves. I have often wondered whether this could be relevant to Miss Frobisher and Twinkle. Whatever the problem, this case was one of the worst examples of a dysfunctional human/cat relationship that I have ever seen. I am so glad that I was able to help Twinkle have one year of freedom.

The positive side

You are probably starting to see that a cat behaviour counsellor's lot is not always a happy one. If ever I drive home emotionally exhausted with tears running down my face it is after seeing a case that relates to a bonding problem. They often seem to result from a deep sadness and, during the consultation, I get an insight into that suffering. I have an ability to absorb emotion like blotting paper and I often return home frayed at the edges. There must be easier ways to earn a living.

Nevertheless, there is an incredibly positive side to the human/cat bond. People achieve great happiness from their relationships with their pets and these four-legged companions can often be as good a reason as any to get up in the morning and keep going when things are tough. I still find it amazing that the pet cat is capable of such incredible diversity and adaptability and uniquely slots into any bizarre situation with the greatest of ease. We get this idea in our heads that our cats love certain things done in particular ways. We are the only ones that truly understand our individual pets and no-one else could ever love them or care for them as well as we can. Maybe in some instances this is true. Bonds between humans and cats can be so strong that continuing after the loss of a companion becomes unthinkable. I have sadly known a number of cats who have died shortly after losing a human friend. Siamese appear particularly susceptible to all-consuming grief and they can easily stop eating and die.

In the main, I am pleased to say, cats are perfectly capable of surviving without us. They are chameleon-like in their attitude to relationships with humans and can adapt to new situations with relative ease. Just as we can be different people in different relationships, cats can be that way too. Some owners can bring out the best in their cats and others the worst. For example, Chester might not have sprayed urine on his owner and Rover (Chapter 3) might not have attacked people if they had been part of a different human/cat relationship.

I shall never forget visiting a delightful lady who was having a bit of a problem with one of her six cats. She was a very attentive and loving owner and her naughty Birman was taking advantage of her compliant nature. It was a fixable problem and I tackled the consultation with good humour and a positive attitude. Whilst she was explaining to me quite how much she loved her cats she illustrated the point by showing me a lever arch file filled with paper. It was a beautifully constructed dossier on all her cats, containing every possible piece of information about their character, medical history, habits, likes and dislikes. She had compiled this instruction manual as a type of insurance policy against the possibility of both herself and her husband's being killed in a plane crash (I couldn't quite ascertain why she felt the likelihood of an air disaster was greater than any other fatal accident). This comprehensive tome was probably the most useless collection of 'facts' that I have ever read. I didn't have the heart to tell her that it was merely a testimony to her unique partnership with her six cats and that, in the unlikely event of her tragic demise, the relationship would die with her.

❀   ❀   ❀

Zulu kept smiling and eating until the day before he died in the dreadful year we lost three of our beloved cats. He was an imposing and plump figure who loved his food so Peter was worried on New Year's Day when he refused his breakfast. He phoned me and said that he was sorry to worry me but he wasn't happy about Zulu. I told him to take him straight away to the emergency surgery but within two hours he was dead. We still don't know what killed him. He had a very low red blood cell count so it is possible that he bled internally but it remains a mystery to this day. Zulu's death started the terrible tension and distress in the household that led to the deaths of Bln and Puddy and also, indirectly, to Lucy's illness. The change in all the cats was dramatic as they tried to adjust to the various members' absence; it was extremely difficult to cope with my grief and watch the cats suffering in their own way as well.

CHAPTER 8

The Elderly and Disabled Cat

Hoppy's Story

TWO MORE ADDITIONS TO THE FAMILY CAME DURING MY TIME with the RSPCA and one of them was Hoppy. A local elderly gentleman had died and left his constant companion without a home. The family did not want the cat so he was brought into the centre for rehoming. Nobody seemed to know how old he was but it was obvious he was no spring chicken. He was called Hoppy because he had been caught in a gin trap as a kitten and half of one of his hind legs was missing. A modern veterinary practice would have removed the remaining section of leg to allow the body to adjust to a tripod shape. Hoppy's remaining stump had created a curved spine and a difficult gait to compensate for the half leg that was a useless appendage.

Hoppy was a large tabby and white cat with a handsome round face but he was desperately unhappy at the centre. He wouldn't eat and we were concerned that he was pining for his owner. I decided to take him home and see how he adjusted to life with our crowd. It was a cold day and the log fire was blazing in the small lounge we referred to as the 'snug'. All five cats, Spooky, Bln, Puddy, Bakewell and Lucy, were sprawled out in front of the fire asleep when Hoppy was brought in and placed on the carpet in front of them. (They say that ignorance is bliss and I certainly wouldn't introduce Hoppy to the others like that now!) All five cats looked up and hissed and Hoppy didn't bat an eyelid. He approached each in turn and sniffed. He then looked around, looked at the fire and hobbled to a position directly in front of the heat. He then lay down and went to sleep and the others followed suit. Hoppy's grieving for his previous owner was over and he became the leader of his new group instantly. There were no further arguments and no disputing his unquestionable right to be top cat.

Hoppy was an amazing character. He was soft and cuddly and loved nothing more than to be picked up and squeezed and kissed. He was also a formidable defender of the other cats and the property. He would sit at the top of the driveway waiting for our neighbour's dog to start chasing one of the cats. When this happened they would purposely steer the dog towards Hoppy, who would stand his ground. As the dog approached, Hoppy would raise both front paws (not easy with only one and a half legs at the back to balance on) and totally redesign its nose with his claws in a series of savage circular motions. We nicknamed him Hoppy-Scissor-Paws for obvious reasons. He knew that Spooky and Lucy were a little nervous outdoors and it was probably no coincidence that he would be found close by if they were ever curled up asleep in the rockery.

Mary Stewart – the cat with two and a half legs

Hoppy's disability was never an issue for him or any of the other cats. I have known many amputees over the years and I have always been delighted by their speedy recoveries and apparent ease in adjusting to three legs. I have seen amputee cats catch mice, run up trees and beat up dogs! One particular cat that springs to mind is a brave little character called Mary Stewart. I had always thought that she was named after the Scottish Queen but her owners, April and Paul, were merely confused about her sex when she was a kitten so gave her a boy's name as well as a girl's. Mary Stewart was born to a farm cat mother that lived in the fields behind April and Paul's home in rural Cornwall. She joined their cat family and became an independent and adventurous adult.

When Mary Stewart was about three years old tragedy struck and she was the victim of a hideous accident. She didn't come in for her dinner that evening but April (unaware of the circumstances) wasn't unduly concerned. However, she started to worry when she didn't return that night or the following morning. Mary Stewart was missing for nearly two days until April heard a plaintive cry at the back door. There she was, dragging herself with her front paws up the step into the kitchen. April was shocked to see that her back half was a bloody mess. Without hesitation she wrapped her in a thick towel and drove straight to her vet. The prognosis was poor. One hind leg was so badly smashed that it required amputating. The other hind leg was devoid of all the flesh and muscle and broken in several places. There seemed to be no alternative but to put dear Mary Stewart out of her misery. I am still not sure what made April and her vet so determined to save that little cat. It certainly was a dramatic decision when they both agreed to amputate the severely damaged leg but try to save the remaining one.

Months and months went by of surgery, dressings, antibiotics and painkillers. Mary Stewart remained peaceful and the perfect patient throughout. April felt dreadful for her suffering but continued to be driven to save her life. Unfortunately the treatment to the remaining leg was failing and it became apparent that the lower half could not be saved. Another conversation took place between April and the vet and a rather unusual decision was made. That afternoon Mary Stewart had the lower half of her remaining back leg surgically removed. She was destined to live the rest of her days with two and a half legs.

This story had become a legend in the local area and, despite hearing about this marvellous little cat, I had never met her. I knew April, a lovely lady who campaigns tirelessly for animal welfare throughout the world, so I telephoned her and asked if I could come and visit Mary Stewart. I eventually travelled to the little cottage in Cornwall on Mary Stewart's sixteenth birthday. I first saw her curled up on a sumptuous blue velvet cushion in a sunny window. I felt rather as if I were being granted an audience by royalty. She greeted me with polite affection and then decided to jump down from her cushion and go and see what delights were available in the kitchen. She moved rather like Hoppy, with a shuffling gait, but April assured me that (just like Hoppy) she could run really quickly when she wanted to. Her coat was in good condition and her weight was perfect. April had always been keen to keep her weight down to avoid unnecessary strain on her stump. The remaining half of her back leg was a short muscular V-shaped limb with hard skin at the base. She used her litter tray just like any other cat and behaved in every other way as normally as a regular amputee. Her front legs and shoulders were particularly muscular. April said that she appeared to suffer from cramp from time to time, but, despite everything, she had lived a long and extremely pampered life since her accident. I fully expect to be helping Mary Stewart celebrate her twentieth birthday in her idyllic home!

The Elderly Cat Survey

It was the introduction of Hoppy into the household that really fuelled my fascination with elderly cats. It was hard to tell, but Hoppy was probably at least twelve years old when he arrived and I was in awe of his composure and understanding of the world. Nothing fazed him, and his ability to manipulate people and situations to his advantage could take your breath away. I started to wonder, having seen so many cats age in a similar way, whether this was a relatively standard pattern that the domestic cat adopted to compensate for the obvious restrictions of old age. Let's face it, if you can't do it yourself any more (or frankly can't be bothered), why not get someone else to do it for you?

I set to work devising a questionnaire that owners of elderly cats could complete to give some insight into the idiosyncrasies of their ageing pets. With the help of Claire Bessant, the talented and extremely knowledgeable Chief Executive of the Feline Advisory Bureau, I compiled a two-page document covering everything from diet to health and lifestyle to relationships. Now all I needed to do was find enough people to fill in the questionnaire to make the data meaningful.

Several cat magazines and local newspapers printed a short piece of editorial and requested readers to write in to take part in the survey. I was delighted that this resulted in letters from one hundred and eighty-seven willing volunteers with loads of stories to tell. I was well on my way to finding out the answers but the survey was soon to get much bigger. I had a telephone call from Celia Haddon, the pet columnist in the Weekend section of the *Telegraph*. She asked if she could write an article and encourage more people to take part in a worthwhile survey. She is a great cat lover herself and the piece she wrote was funny and relevant to anyone living with and loving an elderly cat. The response was incredible. Within days the post office in my local town in Cornwall ground to a halt. Seasonal staff were drafted in to cope with an unprecedented surge of mail, all addressed to me. Peter and I spent weeks opening letters and sending out questionnaires. I received gifts and photographs from all over the world for some months afterwards, all of them without exception praising the intelligence and wisdom of the elderly cat. I was hooked and the whole project overwhelmed my life for some time.

I recall one particular incident whilst processing the letters and filling stamped addressed envelopes that caused some concern. I opened one piece of mail that merely contained a folded self-addressed envelope. Actually, that's not strictly true. To my absolute horror the latter was discovered to contain the last two or three inches of a cat's tail. It was brown in colour, like that of a Siamese or Burmese, and it was relatively dry and well preserved. Now I had a dilemma. I admit to having a vivid imagination and I was thinking 'Is this a cry for help?' or 'Is this a statement of protest to shock me?' and even 'Is there a poor cat out there suffering with the end of its tail missing?' I explained my predicament to my friends at the RSPCA. We all agreed that there was probably a reasonable explanation, but an RSPCA inspector would call at the address and make sure all was well. The following day I had a telephone call from the sender of the mystery tail. She was a lovely lady full of apologies (and quite a bit of laughter). Apparently, as a result of an accident some time ago, her beloved Siamese had required the end of his tail to be removed surgically. The cat was so adored that his owner couldn't bear the thought of any part of him being consigned to a bag of clinical waste. She therefore asked the veterinary surgeon if he would be good enough to return the leftover bit of tail after the surgery. (I must admit I cannot think of many vets who would be happy to do this, particularly as amputated parts can be rather messy.) When she got home she was unsure quite how this trophy should be stored, so she placed it in an old brown envelope to await a decision on its fate. Needless to say it was forgotten, even when the same old brown envelope was used as an SAE for the survey. She just hadn't noticed the tail tip was in there. We had a lovely chat on the telephone and I returned the end of her Siamese cat to its rightful home. I expect she has it mounted now.

Well over a year later I completed collating the survey results. There is still much information to be gleaned from those questionnaires and I will probably be analysing them for many years to come. I felt it was an important subject for a number of reasons, not just personal curiosity about my own ageing group. Cats throughout the world are living longer, have better nutrition and higher standards of veterinary care than ever before, and while we understand their specific requirements with regard to their changing physiology I didn't feel we were taking into consideration any increased emotional or behavioural needs. The study answered my fundamental question, 'Do cats change that much as they get older?'

The final survey consisted of 1,236 cats over the age of twelve throughout the UK. The results clearly indicated that there are some common changes that take place in the elderly cat's behaviour, many of which can be attributed to particular physiological deterioration or disease. The really interesting information, which cannot be explained quite so readily, is the changing relationship between owner and cat and between cat and its animal companion. Hundreds of letters were received accompanying the completed questionnaires and these held some fascinating insights into the potential bond between human and cat.

The table below shows the demographic results of the Elderly Cat Survey. The oldest cat in the survey was Stevie, who was 26 years old at the time.

12–15	56%
16–19	38%
20+	6%
Neutered males	45%
Neutered females	55%
Entire cats (male and female)	<1%
Owned from a kitten	68%
Adopted as an adult	32%
Domestic shorthair	74%
Domestic longhair	7%
Siamese	7%
Burmese	6%
All other pedigree breeds	6%

Nutrition

The first category the owners were requested to give information on was nutrition. Fifty-six per cent of those surveyed said that their cat's appetite had stayed the same, with 20 per cent saying it had increased and 24 per cent decreased. There are various physiological changes that occur in old age that can account for increases and decreases in the amount of food consumed. Decreased sense of smell, vision and ability to taste will reduce the food intake because of the importance of those senses in appetite stimulation. Dental problems can also be a consideration in older cats; periodontal disease affects 85 per cent of all cats to one extent or another and this can dramatically affect the appetite if the problem remains unresolved for any length of time. Often dental disease will be present long before the cat actually goes off his or her food.

A general reduction in metabolic activity and exercise means the older cat will require less food and there is often a genuine need to reduce the calorific intake to avoid obesity. Obesity can lead to medical conditions such as diabetes and this can be a difficult condition to control. Other diseases such as hyperthyroidism can increase the appetite, but on the whole it appears that the majority of older cats maintain similar appetites throughout their lives until illness interferes.

Half of the owners surveyed had been 'trained' to feed their cat on demand. Only 1 per cent fed once a day, whilst an equal proportion fed their cats two or three times a day. A cat left to its own devices would choose to eat little and often and it is likely that cats fed periodically would return to the bowl several times during the course of the day. Feeding time is an opportunity for loving interaction between cat and owner. There is an innate feeling in most people that a good appetite is directly related to good health and this belief has probably played a role in the increasing incidence of obesity in our pet cats. There is also the philosophy 'I love therefore I feed' in many owners. This concept was very apparent in many of the letters received accompanying the completed questionnaires. Here are a couple of extracts describing the importance of food in the owner/cat relationship:

Sam doesn't eat quite so much now, he seems to have a little at a time, but there is always food down for him. I cover it over with a saucer; when he wants more he knocks it off with his paw. (Seventeen-and-a-half-year-old domestic shorthair)

Ginger does demand food more often but I don't think it's an increased appetite as much as the chance to have an extra bit of pleasant interaction with me. (Thirteen-year-old domestic shorthair)

Whilst appetite may not increase dramatically in old age it was apparent from the letters that many cats devote more time to acquiring food even if the quantity consumed doesn't change. Cats really do have a unique ability to train humans and they soon learn that hovering in the kitchen miaowing results in the production of tasty titbits. As they get older, cats are likely to be spending more time at home and this inevitably leads to more opportunities to feed. In a letter received from a lady in Oxfordshire, whose cat Smudge is seventeen years old, she actually refers to this behaviour to illustrate her cat's deafness: *We noticed she was not responding to calls or the fridge door being opened.*

The results showed a fifty/fifty split to the question, 'Has your cat become fussier about food as he/she has got older?' Many cats had developed fussy appetites after being offered a large variety of alternative foods, all slightly more palatable than the last. Often a temporary 'loss of appetite' is sufficient to cause the owner to provide even tastier treats. This appears to be manipulative and opportunist behaviour that many cats learn, not just the old ones (remember Chapter 7?). Owners will frequently offer food that they are eating themselves. Whilst not all titbits are exciting, they are certainly worth investigating just in case. It appears also to be very comforting to have a cat enjoy human food, creating even more in common between us and the 'little furry humans'. There was a tremendous sense from the owners' letters that they feel the pleasures available to the elderly cat are limited and food is perceived as an important part of their lives. This goes a long way towards explaining how the aged cat can become 'fussy'. After all there is no incentive to eat tinned food if there is smoked salmon in the fridge.

I found the following quotes very telling:

Her favourite is tuna and mayonnaise paste. She is very fond of yoghurt which she will eat from the carton.

He also enjoyed the same breakfast cereal as my husband and I.

At first she would eat nothing but cauliflower cheese and fresh green peas.

They also enjoy garlic, coconut, veggie cheese etc.

Smokey who lived to be twenty loved orange peel.

I am constantly amazed how many cats seem to survive on totally inappropriate diets. I insist nobody tries the 'cauliflower cheese and peas' diet or any other vegetarian equivalent because this would cause serious illness. Cats are obligate carnivores and they need an animal source of protein in their diet. Obviously some of these cats were experimenting wildly with new tastes but I am pleased to report that all the above examples related to cats fed these items as treats, not as a staple diet.

There is no substitute for a good quality diet specifically formulated for cats. However, if a cat loses its appetite in the latter stages of terminal illness or is generally failing as a result of old age, it is often possible to feed tempting morsels and give a decent quality of life at the end.

Sleep

The major pastime for the elderly cat is sleep. This includes everything from deep sleep to cat naps and resting with shut eyes. Forty per cent of the cats in the survey slept for more than eighteen hours a day. Most of these cats were in the 16–19 and 20+ age groups, which certainly would be in line with the general ageing processes and the slowing of the metabolism. The majority (57 per cent) slept for twelve to eighteen hours a day and only 3 per cent of those surveyed slept for less than twelve hours. Most owners reported that their cats certainly slept more now they were old. Remember, elderly cats are going out less, exploring less and generally doing less – enabling them to fill those voids with more opportunities for rest and sleep.

Seventy-eight per cent of owners said that their cats had a favourite place to sleep and it was not surprising to see that almost all the places chosen were near to a source of heat. As a cat ages, its ability to regulate its body temperature is reduced and an older cat is more prone to hypothermia and generally feeling the cold. It is also more likely to seek out a place that is soft, since loss of weight will lead to bony prominences which can easily become sore if pressed on hard surfaces for any length of time.

Top ten favourite sleeping places

1   Owner's bed (45%)
2   Armchair (26%)
3   Outside in the summer sun (11%)
4   Near the radiator or in a hammock
5   Cat igloo/bed/basket
6   Airing cupboard
7   Owner's lap
8   Conservatory/greenhouse/sun lounge
9   Anywhere in the sun indoors
10 Near the Aga/boiler

All of the above places are warm or soft or both. If you are spending most of your life asleep it is important to put effort into finding the best place to do it.

Territorial and outside activities

Four per cent of all the cats surveyed had lived exclusively indoors. Of the remainder, 55 per cent were going out less than they used to. Thirty-nine per cent were going out about the same amount but these tended to be in the 12–15 age group; most of those who were sixteen and over were finding home comforts more attractive. Only 6 per cent were going out more often than before; these tended to be cats from multi-cat households who were reported in their old age to be less sociable and more remote with the other cats. This would probably be an avoidance strategy rather than a desire for the great outdoors.

The time spent in the younger years hunting and patrolling territories and being out in the cold and wet are, given the choice, almost bound to reduce in the older cat since thermoregulation and general mobility are declining. Hunting is inevitably going to suffer. Visual and auditory senses are dimming and arthritic joints are not conducive to successful stalking, chasing and pouncing

Still hunted	31%
Never hunted	22%
Stopped completely	47%

A third of the older cats surveyed were equally aggressive as when they were young in defence of their territory against outsiders. These cats are probably maintaining the habits of old, since it is unlikely that they are coming off worse time and time again; their challengers are probably not calling their bluff. Another third were more tolerant, choosing the 'live and let live' way of life in their old age. This does seem the sensible option of the elderly, since it is doubtful whether any fight would be won on sheer strength and mobility. The remaining third avoided conflict by running away, ignoring the outsider or glaring from behind the safety of a window.

Cat companionship

The next section in the questionnaire related exclusively to those cats living in multi-cat households; a total of 59 per cent of those surveyed. Just over half had remained the same towards their companions in their geriatric years. It appears that cats vary as they get older. Some mellow, some become cantankerous, some actively seek the company of other cats. The less tolerant and more remote were quite often those in households with kittens or young cats. Their constant movement and play is not helpful to an old cat who just wants some peaceful uninterrupted sleep, so the idea that a new kitten will give an old cat a new lease of life is not always the case.

By far the largest number of letters received related to the reactions of elderly pets to the death of a cat friend. Forty-seven per cent of the cats surveyed had outlived another and 60 per cent of those showed some visible reaction to the loss. The Siamese, Burmese and Birman were particularly well represented in this section. Some owners told of their cats becoming more affectionate and demanding, some even said that the cat improved tremendously and appeared more content since the loss of its companion, but almost all the reactions reported included searching and calling. Here are a few examples.

Solomon was very disorientated and confused and obviously very insecure. He looked for Cy around the house and called a lot. This lasted 2–3 weeks. (Eighteen-year-old Birman)

From that day on Kula began to grieve for Kiki. She cried day and night for her. There wasn't a cupboard, drawer or wardrobe she didn't look in or pull things out of to see if she could find her. (Twelve-year-old Siamese)

He howled day and night for months and eventually started to wet and soil on the carpets. He was cured more or less instantly by the acquisition of a new blue kitten which he accepted immediately. (Thirteen-year-old Burmese)

He became profoundly deaf after his brother died. He no longer purrs nor does he approach with his tail elevated. He seldom washes, just his face after eating, and when he's awake he stalks around the house making a horrible sort of calling miaow – quite unlike the greeting sounds. It is difficult not to attribute these changes in personality to a type of bereavement syndrome. (Nineteen-year-old domestic shorthair)

Snowy had to be put to sleep because he lost the use of his legs. Vickey died later the same day – it was very sad. (Twenty-one-year-old Siamese)

When Biz died it took her sister about five days to realize it and then she started to howl day and night. Since the kittens have been with us the howling has stopped. (Eighteen-year-old domestic shorthair)

This subject seemed to grip the owners surveyed. It is interesting to note the number of cats who improved tremendously when new kittens were introduced. I do not feel this is behaviour unique to old age. Similar instances have been reported in younger cats. The relevance of age is that the companions have often been together for a very long time and the desire for routine and lack of change appears to be heightened in the elderly. The loss of a long-term friend creates a profound difference in the household – grieving humans, changes of routine and the absence of a familiar part of the domestic unit – and probably accounts for the distressed calls and searching. The cat wants things to return to normal. The introduction of a kitten is sometimes the trigger which stops the unsettling behaviour by occupying the mind with a new source of company. There is also the other side to the coin, with those owners who reported that the remaining cat 'blossomed' on the demise of the other. It appears that passive oppression between cats may only become apparent when the assertive one is no longer there. The survivor can develop a more confident and friendly nature and start to sleep in the dead cat's favoured resting places. Is this a mark of respect or the symbolic claiming of the rank of top cat? Probably the latter!

The Orientals are bred to be very devoted and loving towards their owners and they are often very sensitive to change and moods. It is quite understandable that they would become distressed at losing a cat companion if they bond in the same way with other cats. Whether it can truly be evidence of a grieving process as we understand it is debatable. When we lose a loved pet it is helpful in dealing with our own grief to feel that a remaining cat has an empathy with our sadness, but, in behavioural terms, are we merely seeing a withdrawal response to an addictive relationship that has abruptly ended?

The owner/cat relationship

Almost without exception, the elderly cat turns to us for love and attention in their old age. Eighty-one per cent of the owners surveyed reported that their oldies had become more sociable and affectionate or more demanding of attention or both. Only 2 per cent felt they were less sociable, whilst the remaining 17 per cent stated they had always been very affectionate or independent and really hadn't changed at all. A number of people had experienced a tremendous change after a period of illness, resulting in a much more dependent and loving cat.

Vocalization appears to play a big part in the ageing process. Sixty-six per cent of the cats surveyed use more sounds to get food and attention. For example, a thirteen-year-old Siamese crossbreed *howls and demands food*. Josie *has a loud harsh miaow which she did not have when she was young*. A twenty-three-year-old domestic shorthair *is constantly miaowing*. A thirteen-year-old domestic shorthair *makes sounds much more like a kitten squeaking than the miaowing you'd associate with an adult cat. Her vocal range is wider than it used to be. She now has lots of different miaows and other sounds to indicate different needs and wants.*

As cats get older, their owners become more and more in tune with their needs and cats soon learn to play on this. It is not unusual for a variety of noises to be used if they result in attention, affection or food. Only 4 per cent of those surveyed said their cats called less. The remaining 30 per cent felt they were calling about the same as they always had done.

Night-time vocalization

A particularly disturbing habit of the elderly relates to harsh vocalization at night. Twenty-eight per cent of those surveyed called for attention at night and stopped only when they received attention or reassurance from their owners. Of these 346 cats, 54 per cent started the behaviour between the ages of ten and fifteen. As a cat's ability to protect itself declines there appears to be a higher dependency on its owner for security. Maybe these howling cats, having enjoyed additional attention during the day, feel in need of reassurance when their owners are not around in the night? Having tried successfully a number of times to elicit a response from their owners (this is a harsh distressed yowl that is difficult to ignore) they continue to perform the ritual as a habit. A number of owners reported that the calling stopped when the cat was allowed to sleep in the bedroom. However, the cat will often jump off the bed and wander off downstairs only to repeat the crying all over again.

There are a number of physical reasons why this calling may occur. Deafness seems to play a role in the harshness of the cry (I can't imagine how this sound feels in the head of a deaf cat). It is possible that a chronic deficiency in the oxygen supply to the brain could possibly produce symptoms of senility and short-term memory problems causing confusion at night. Cognitive dysfunction may cause a change in the sleep/wake cycle resulting in some cats' being awake at night at times when they had previously been asleep. High blood pressure causing general discomfort, headache and disorientation could also promote this sort of distress response.

Night-time vocalization is often reported as one of the behavioural symptoms in cats suffering from hyperthyroidism, a condition seen frequently in the elderly cat. A tumour on the thyroid gland causes metabolic changes including increased heart and respiration rates, increased appetite and weight loss. I have had three cats with this condition, Bln, Hoppy and Bakewell, all of whom cried pitifully at night.

Play

Other changes in habits and behaviour can be observed when the owner interacts with the elderly cat. The play response is still there for some but this mostly has to be instigated by the owner. The general deterioration of joints and mental agility makes fast turns and rapid movements less possible. Only 10 per cent of owners said their cats still played regularly, 48 per cent said occasionally and 15 per cent said they had stopped completely. The remainder had never played with their cats. I think this is a shame. A cat should be encouraged to play in its elderly years to provide exercise and stimulation. The games may not be quite so boisterous as before but will certainly be beneficial for both cat and owner. Both Spooky and Hoppy loved nothing more than batting a feather on a string with one front paw whilst reclining on a soft cushion!

Grooming

Grooming habits are affected by ageing, since stiffness reduces the suppleness necessary to do the job thoroughly. You will often see old cats with mats towards the base of their spine and this is undoubtedly due to the fact that they can no longer flex their body sufficiently to perform an efficient and thorough grooming regime. The frequency of washing may not alter until very old age but it is almost certain that areas will be missed. Three-quarters of the cats in the survey still groomed regularly, 22 per cent occasionally and 2 per cent had stopped completely. Chronic illnesses and toilet accidents were also reported in this last group and this is in line with the idea that the very sick and elderly will not groom. Most elderly cats benefit greatly from combing and brushing from their owners, if care is taken about the prominence of the bones in skinny cats and the discomfort a harsh comb would cause.

Toilet habits

Litter trays were provided for the elderly cat by 55 per cent of owners. The other 45 per cent reported that their cats chose to go outside. Twenty-nine per cent of cats had toilet accidents since they had become elderly. A number of owners related these accidents to illness, for example cystitis, a bout of diarrhoea or even the development of incontinence in the very elderly. Many older cats start to have 'accidents' indoors and this is usually found to be a result of an increasing reluctance to urinate and defecate outdoors, due either to the presence of aggressive cats in the territory or an increased sensitivity to inclement weather conditions. The provision of an indoor litter tray almost invariably solves the problem.

Percy – house-soiling in the elderly cat

I remember with a great degree of affection a certain dignified old 'gentleman' called Percy. He was seventeen years old when I first met him; a handsome if somewhat tatty long-haired black and white moggy. He greeted me when I visited with a casual flop and a welcoming purr. He was delightful and his owner Joan was equally friendly (though without the flop and the purr).

She had telephoned me about a distressing problem that had started over the past few months. Percy, after seventeen long years of toileting outside, had suddenly taken to urinating and defecating in several specific locations throughout the ground floor of their lovely house. Joan and the family were devastated and automatically presumed he must be poorly. Percy had previously been diagnosed with chronic renal failure and his kidneys were still struggling a little. This made him drink and pee rather a lot but their vet felt that the condition was relatively under control. He doubted that it would suddenly account for this change to his normal habits and he suggested that Joan speak to me.

Percy and his family lived in a Victorian terraced house in north London and he had always enjoyed free access to the great outdoors via a cat flap. Joan explained that Percy had been the perfect pet all his life. He and his sister Portia had grown up with the family, and since her death two years before Percy had muddled on without her. He used to go out every morning, jump over the fence, then wander down the passageway at the back of the garden to carry out his secret assignations. He had followed this routine most of his adult life but there were a large number of cats in the area these days and Joan said that Percy had been beaten up outside rather a lot recently.

A few months before my visit Joan had extensive building work done and new laminate flooring put down in the sitting room. The place looked great but unfortunately the improvements had coincided with Percy's pools and piles, deposited on a regular basis in four or five private corners in the kitchen, hallway and sitting room. Joan was appalled and she hated cleaning it up. Her husband used to call Joan every time he discovered something with a bellowed 'Percy's left a little present for you, darling!' (Oh, well. At least he was fairly laid back about it.)

The worst thing as far as Joan was concerned was that Percy was obviously unhappy. He wasn't grooming as much as he used to and he just looked depressed. She didn't blame him at all; she just desperately wanted him to feel better.

I explained to Joan that as cats get older they become less confident in their ability to cope with day to day challenges. Any changes in routine or lifestyle, even the different flooring in the sitting room, can be extremely distressing. Percy had been through rather a lot recently. His companion had passed away, he had been beaten up, the kitchen had changed beyond belief and the floor in his favourite room was now so unbelievably slippery that he distrusted it immensely. I also suggested that, if we got inside Percy's head, we would probably find other complicating factors. What if Percy was thinking (in his own feline way) 'I swear that fence at the back of the garden has got bigger. I just can't get over it as well as I used to' or 'I am so stiff and tired and it's freezing out there; I just cannot bear the thought of going all that way just for a wee' or even 'What if that black cat jumps on me again? I can't take many more beatings!' We both agreed that things had to change. Joan had to prepare herself for a shock. Percy, that lovely black and white kitten who used to give the kids so much pleasure, was now an old man. He couldn't do the things he used to and it was up to the family to be considerate and sympathetic and to provide him with the creature comforts that all elderly cats need and deserve. Percy was after all the equivalent of eighty-four years old.[1]

Joan approached the new regime with enormous enthusiasm. Percy was provided with two shallow and discreet

---

1 The formula for estimating age in cats is as follows: the first two years are equivalent to twenty-four human years. Thereafter it is calculated at one cat year to four human years.

litter trays in private areas in the kitchen and under the stairs. He had four small meals a day (instead of his usual two) and each meal was accompanied with love and praise. Every day Joan groomed him and his coat became glossy and tangle free. Percy was sleeping a lot because of his age and several new thermal beds were positioned throughout the two floors to ensure that a cosy place could be found anywhere in the house. Joan covered a section of the new laminate floor with Percy's very own rug and he took to it straight away. He actually appeared glued to it for several weeks; Joan thought he was worried they would take it away when he wasn't looking. Joan was also instructed, as part of the programme, to play with Percy gently every day. She made a fishing rod toy with two feathers stuck on the end and Percy used to fling himself around (as best one can at eighty-four) to try to catch it. Dear Percy never soiled again and Joan felt he had a new lease of life.

Two years later Joan called me. I remember I was lost in the back streets of Portsmouth at the time (looking for a client's address) and I had to pull over to hear her news. Percy wasn't well. He had started eating his cat litter and generally behaving strangely. We had a chat and I advised her to visit her vet and get some blood tests done just in case there was an obvious problem that should be addressed. I talked about senile changes and generally waffled on and hoped that I had given her a little hope. She didn't call again but I thought about him a great deal over the next few weeks. In my job you have to remain slightly detached. If I were to keep in touch and make unsolicited contact with previous clients I wouldn't be able to maintain standards for current patients. It's frustrating but I only have one brain and that has a finite capacity! I would have loved to call Joan afterwards and find out what happened to

Percy but I knew I had to hold back. It's a safety precaution, really, for my own sanity. However, some months later, at Christmas, I had a card and a letter from Joan. Percy had been peacefully put to sleep in her arms at the age of nineteen and a half. They had buried him in the garden next to Portia. She had wanted to grieve for a decent period before getting another cat but apparently fate had intervened and two kittens had been found that were desperately in need of a good home. Joan said that she felt that Percy wouldn't want her to waste her cat love. I agree, Joan. Those kittens are very lucky!

Senility

Many owners in the survey spoke of a number of character changes and unusual behaviour which they had, possibly quite correctly, interpreted as senility. A blank expression, getting lost in familiar surroundings, constant yowling, lack of grooming, continuous pacing, inappropriate toileting; all with no obvious physical cause. There appear to be uncanny similarities between the symptoms shown by the elderly cat and those of a human dementia patient.

One lady described her twenty-year-old cat's behaviour in a way that paints a vivid picture: . . . *paces round and round, starts in the dining room, straight through to the lounge, along under the front window, then back to the dining room. She will do this endlessly. Sometimes there will be a slight change: while walking past she will jump on your lap, just walk over you, jump down then carry on pacing.*

Milly – the senile cat

Milly was a sweet but decrepit and fragile-looking creature of indeterminate age. Her owner, Bernadette, had become increasingly frustrated with Milly over the previous year as her behaviour had started to get rather strange. She had taken to standing or sitting in front of the cooker and staring at it for long periods at a time. This had become extremely inconvenient if she decided to do so when Bernadette was trying to cook dinner. She had also developed a howling wail at night that kept not only her owners awake but the neighbours on either side of their terrace house as well. Bernadette would often get up in the night in response to this terrible cry and find Milly sitting in the living room staring at the wall. Poor Bernadette was utterly frustrated by this change in her cat's personality. She felt that Milly always had a look of distraction on her face; she tried to explain it by saying, 'You know that look that says the wheel's still turning but the hamster died long ago!' I told her I knew exactly what she meant and tried very hard to help both owner and cat adjust to the possibility that there were senile changes taking place in Milly's brain.

Thorough veterinary investigation is always essential when exploring the reasons for abnormal behaviour, particularly in the elderly. We had previously ruled out possible medical causes such as hyperthyroidism (a tumour on the thyroid gland), high blood pressure as a result of kidney failure, and a number of other potential diseases. The referring vet found nothing particularly untoward so we were left with an age-related behaviour that needed understanding and compassion from the owners. Milly was becoming forgetful, absent minded, insecure and frightened. This was particularly evident at night when the house was quiet and she would wake suddenly from sleep and find herself alone in the dark. She would call in distress, probably trying hard to work out where she was, and her owner would come running and provide comfort and reassurance. Milly knew that, if she did feel this way, at least her owner would be there for moral support.

We had to improve Bernadette's understanding of Milly's condition and the whole night-time thing had to stop. Whilst we all want to comfort our cats at times of distress it is unrealistic to feel that we should be 'on call' twenty-four hours a day. We had to devise a plan that gave Milly a better sense of security at night. Based on the principle that she was getting confused by her nocturnal wanderings, I felt that she would benefit from a small 'core area' that provided comfort and predictability for a good night's rest. The kitchen was a cosy room and a particular favourite of Milly's. I therefore suggested that a bed be created adjacent to the radiator and that the room be set up to provide all her needs including litter tray, food, water etc. I made Milly a bed out of a cardboard box with the front cut out for ease of entry and exit. This provided protection from draughts and a snug place to curl up. It was lined with a thick pillow that was covered with a piece of thermal synthetic sheepskin bedding. (This stuff is available from your vet or good pet shops and makes a great washable bed for the elderly.) I advised Bernadette to feed Milly small frequent meals and, just before she went to bed, to take Milly into her new 'bedroom' and give her a loving cuddle. She would then be shut into the kitchen overnight. Any calling would be ignored until the following morning.

Bernadette was sceptical about a programme that appeared

to consist merely of a cardboard box! There were actually lots of other little things being done, for example a new play and grooming regime, but the cardboard box certainly proved to be the most popular addition to Milly's life. Bernadette informed me that she had heard Milly crying the first night but she had followed instructions and remained in bed. Miraculously this was the last time Milly cried out at night. She appeared to be extremely grateful for her cosy new bed and she spent a great deal of time with her tiny head leaning on the edge of the box surveying her domain. She started to enjoy fresh air in the secure garden outside and even used to pat a toy mouse when she felt in the mood. Bernadette described living with her as 'caring for an aged relative' and all her impatience at Milly's little foibles disappeared. Bernadette tried to keep a strict routine within the household and Milly spent her last remaining year in a comfortable and loving environment.

Age-related illness

In the latter part of the questionnaire I asked whether these old cats were suffering from any specific illnesses. As discussed before, chronic illness in old age is a factor that can affect behaviour. For example, kidney problems will make the cat drink more and deafness will make the cat unresponsive and more vocal. Thirty-eight per cent of cats surveyed were suffering from chronic or terminal illness. The most common (according to the comments from the owners themselves rather than from a veterinary source) are, in descending order:

Arthritis
Chronic renal failure

Deafness
Blindness
Hyperthyroidism
Bronchitis
Dental problems

The number one condition was an unexpected winner. It is only recently that veterinary pharmaceutical companies have started to acknowledge that arthritis is a major problem in elderly cats. Painkillers are now being produced specifically to treat this problem.

Living with an elderly cat

When I asked in the questionnaire, 'Is your cat as much pleasure to you now he/she is older?' 97 per cent of all the owners said yes. Most replied that their cat was one of the family, giving so much for so little. Many owners took great pleasure from the fact that their elderly cats needed them.

After twelve and a half years he's part of us, like an old child . . . he's really no bother at all and we get so much in return.

During the ten years we have had her not a day has gone by when she hasn't shown her love and thanks to us.

A cat's behaviour depends on the owner giving companionship and activating the cat e.g. by carrying outside on a sunny day and by talking joyfully.

I adore her – she adores me and we cuddle frequently.

She can be very funny and knows it.

A very loving lap cat – very dependent and grateful.

Only 3 per cent said their cat was no longer a pleasure, *more demanding and not so interesting* or *always sleeping*. This feeling represented a very small percentage of owners and I hope for their sake that anyone caring for them in their old age does not have such high expectations.

Over fifty survey questionnaires were completed about cats after their death. It is not surprising to see what an enormous part an elderly cat plays in a family and the tremendous grief experienced when they are gone.

Despite the vet's best efforts she died in my arms after one of the most awful afternoons I can remember.

We hope that soon we will be able to laugh at all the little things she did and thank her for all the years of pleasure she has given us.

Quite often these cats will have grown up with the children of the family, or represent a link with a dead partner or happier times. When the cat dies the association with the past or the particular loved one dies with it and this can create what appears to be an insurmountable feeling of grief.

Caring for the elderly cat

As you can see, there is great variation in a cat's response to age due to genetic, dietary and many other considerations. Some cats look older at ten than others do at twenty. The only conclusion that can be drawn from this survey (as if we didn't already know) is that every cat is an individual and that doesn't change as they get older. Most obvious behaviour patterns in the elderly cat have their roots in physiological changes and the general ageing process. Others relate to the cat's incredible ability to train humans!

It is possible, however, to offer general advice to the owners of elderly cats, specifically geared to providing a sympathetic and caring home in their twilight years.

- Ensure that your cat receives annual check-ups at your veterinary practice. Some vets advise six-monthly examinations for the elderly.
- Elderly cats often benefit from frequent small meals.
- Your cat may need his or her claws trimmed. They cannot be easily retracted in the elderly cat and they can get caught and hamper movement around the house.
- Provide a number of warm, soft and quiet resting places for your cat to spend a significant proportion of its time. If these places are high then care should be taken to offer a number of steps up to assist arthritic joints.
- Continue to stimulate your elderly cat's mental agility with gentle games.
- Your cat may turn to you more for comfort and reassurance. Humour it and enjoy the extra love and affection!
- Groom your cat regularly using soft brushes and combs, particularly around the base of the spine and other areas that are no longer accessible. Take care to avoid areas where the bones are prominent as this can be painful.
- Consider a new animal addition to the family carefully before going ahead. If an old cat appears distressed or lonely

following the death of a companion allow a reasonable period to elapse before considering a replacement. It may be anxiety as a result of the change in routine rather than a genuine loneliness.

- Routine is extremely important, particularly at times when family members are away from home. Friends or professional house-sitters should be employed to care for the elderly cat in its own home to avoid the distress of a change in environment, unless the cat has always been used to frequent cattery visits.
- Provide indoor litter facilities, particularly if there is any suspicion that the elderly cat is being bullied outside or appears reluctant to go out in bad weather.
- For the very elderly, whose world appears to reduce in size the older they get, provide a bed, food, water and litter facilities in reasonable proximity to each other so that they are all easily accessible. (This is probably the only exception to the general rule about keeping toilet and eating arrangements apart.)
- Seek veterinary advice for harsh night-time vocalization.

There are obviously many other considerations which will depend on the individual cat. There are several good sources of information about the elderly cat including veterinary practices, breeders, cat charities (Cats Protection and the Feline Advisory Bureau) and pet behaviour counsellors. The important thing is to remain vigilant and to understand that most elderly cats are non-complaining and stoical, even when they are feeling bad. I think they deserve to be as comfortable and content as possible, don't you?

❖   ❖   ❖

Hoppy became ill shortly after Spooky died; they were very close and I think stress brought on his condition. He developed hyperthyroidism but it was successfully resolved through surgery. Unfortunately there were obviously other problems and Hoppy went blind suddenly after he had been with us about five years. It is hard for us to imagine the enormous impact of a sudden loss of sight and I was concerned that his quality of life would be greatly impaired. I really shouldn't have worried because he coped extremely well and never lost his status in the group. A cat's senses are so acute in comparison to ours that they can easily compensate for the loss of sight or hearing. Hoppy, like every other blind cat, navigated his way around the house using sounds, vibrations and his highly efficient and resplendent whiskers. This wonderful radar system allowed him to detect objects and narrow openings and enabled him to hobble around just as he always did. If you ever live with a cat that becomes blind through accident or illness then don't despair. Just remember these few golden rules:

- Blind cats can continue to have a good quality of life so don't feel this is the end!
- Try to maintain continuity in the home and try to avoid moving furniture.
- Be aware that new objects and unfamiliar smells will have a greater impact on your blind cat so try to introduce new challenges gradually.
- Continue to play with your cat – he or she will still be able to chase toys by using other senses apart from sight.
- His or her whiskers are now vital for orientation so ensure they are never cut or chewed by a cat pal (some cats love to chew their companion's whiskers in overzealous grooming frenzies).

Hoppy became quite a local celebrity and a favourite with many friends and visitors. His photograph had been featured in the *Daily Telegraph* and the Cornish newspapers before his illness and I received many letters praising his good looks and dignified expression. I didn't like to tell people that the confident stare directed at the camera was actually focused on a piece of ham being dangled above the photographer's head!

We had the benefit of Hoppy's company for six wonderful years until one day he suffered a massive stroke and died. It was a tremendous shock for everyone and it took the whole family, feline and human, a long time to come to terms with his death. We all lost a big character that day; he will never be replaced.

CHAPTER 9

Coping with Bereavement

Puddy's Story

SOME YEARS AGO, I WAS APPROACHED BY A COUPLE OF colleagues who ran thriving pet behaviour practices in the southeast to see whether I would be willing to take their cat cases and start a referral service in that part of the world. The timing seemed right somehow, and I agreed. When I had to leave the cats to move back to Kent I hated the thought that they would miss me. I had often counselled people about the need to selflessly do what was right for their pets and I knew there was no way I could preach one thing and practise another. I could never offer Puddy anything like the lifestyle she already had in Cornwall. I knew that I would have to live in a modest flat and there was no way that I could expect Puddy to adapt. It is perfectly acceptable to go from a flat in Kent to a rural retreat in Cornwall but certainly not the other way round! At the end of the day all that I could offer her was me and that just wasn't enough.

As I waved a tearful *au revoir* (never goodbye) I worried more than anything that my lovely cats wouldn't recognize me when I returned for a visit. I really shouldn't have worried where Puddy was concerned. I used to go home as frequently as possible and when she saw me her little face would light up. Immediately she would turn and push her bottom towards me for a good slapping! I don't know how I first discovered she liked to be slapped around the back end. It's not the sort of thing that you would naturally experiment with – 'I know, I'll just give my cat a good slap and see if she likes it!' – so I think it just progressed from normal stroking and petting and an increasing ability to understand what she wanted. A good firm rub and a gentle slap from side to side so that she fell into each of my hands, and Puddy was in heaven. I've since found out that many cats enjoy a similar rather masochistic loving. Strange creatures.

When Puddy was a kitten we used to worry that she was mute. She would rush up to me, look into my eyes and open her little mouth and go, ' ' Absolutely nothing. Puddy was undoubtedly a skilled exponent of the silent miaow. This really was an endearing quality but I secretly hoped that one day she would say her 'first word' (notice a little anthropomorphism creeping in here?). I had to wait until her second birthday but it was worth it. I clearly remember opening the fridge one day (usually a good way of getting all the cats to come from various directions) and finding myself surrounded by four of them, all apparently starving. A discordant round of dubious harmonies began, all of them competing to make the biggest impression. Suddenly a silence fell and three of the quartet turned to Puddy as she took a deep breath. With a serious expression of great concentration she opened her mouth and went, 'a'. She jumped back, as surprised as the others that something had actually come out. The 'mi' and the 'ow' remained missing for the rest of her life but I didn't care. Puddy and I could at last have some meaningful conversations and converse we did as she made up for lost time and became a real chatterbox. She didn't stop 'talking' until shortly before she died.

On 27 February I took my darling Puddy to the vet because she had started to dribble slightly and she was reluctant to let me open and examine her mouth. Imagining nothing more sinister than an infected tooth I was shocked to be told that she had an invasive malignant oral tumour. I tried to be very sensible and professional when confronted with this news and discussed any possible options for treatment. The vet felt that using steroids and antibiotics to fight the associated infection would alleviate her pain but surgery or curative treatment was not an option. Puddy was sent home for love and observation until such time as she and I were ready to call it a day. I felt utter shock and disbelief that, still reeling from the loss of Zulu and Bln over the previous three months, I now had to face the imminent death of my best friend. Driving back in the car from the vet's, Puddy looked up at me and said 'a' and I had a real sense that she was comforting me. Nonsense, I know, but your mind finds ways to get solace in dreadful situations. The real comfort for me was an understanding that cats have no knowledge of their mortality. They know when they are in pain and they probably know when the end is near but they are resigned to its inevitability. I promised myself at that moment to put my selfish emotions to one side and concentrate on ensuring that

Puddy would not suffer and that I would end it when *she* was ready.

We faced the final dreadful moments too often that year but somehow both Peter and I coped by focusing on the well-being of our beloved cats and trying to objectively assess their quality of life at the end. One of the most difficult decisions a pet owner will ever have to make is when the time is right for euthanasia. As a veterinary nurse and cat behaviour therapist I have counselled owners about the importance of letting go to prevent suffering. Everyone has individual views on this emotive subject but it is essential that an owner is able to make an informed decision. Veterinary surgeons and veterinary nurses see an awful lot of death and suffering in their job. It isn't all *Animal Hospital* and healing. There is always the inevitable end and sometimes we have to face situations where pets are allowed to live and suffer for too long. This has had a profound effect on me and has greatly influenced my strong views about my own pets. They have given me so much love and affection and loyalty over the years. The biggest gift of love I can give to them is to prevent terminal suffering and gently and compassionately put them to sleep. We have no right to be selfish at this time.

I am lucky enough to be an informed owner who under-stands my pets' physiology and emotional states. Most owners, however, would not necessarily know how to assess the well-being of their gravely ill pet. I would always recommend being guided by the veterinary surgeon in such matters, but the paragraphs below will show you a way that you can be actively involved in the decision. This may not be every owner's idea of a good thing. Many would prefer not to take such a proactive approach to deciding the timing of their beloved pet's death. I truly respect the fact that everyone is different and there is no easy way to cope with the end of your cat's life. It will always be painful and wretched but I honestly believe that our feelings should be secondary at such a time. Think of all the incredible pleasure our pets have given us over the years. Don't we owe it to them to put them first?

Pain and suffering

Determining the presence of pain in cats is notoriously difficult, particularly when assessing chronic pain or that associated with disease. A cat's behaviour or normal routine can be affected in many ways if pain is present; some of these changes are obvious (think of Muffin in Chapter 3), whilst others are far subtler. Skeletal or muscular pain will affect the animal's ability or desire to climb, jump and play and may even cause lameness or an unusual gait. The only thing that you may notice could be the fact that your cat doesn't jump onto his favourite window sill any more. Often cats will be less active, filling the time with increased sleep. An understanding of the individual is as important as recognizing normal feline behaviour when assessing pain so it is always worth listening to your 'gut feeling' that all is not right.

It is very difficult to view a cat at any one moment and see obvious signs of pain. This is where the 'activity budget' that I described in Chapter 4 can be so useful. Changes to a cat's normal routines and activities are potential indicators that things are deteriorating. If activity budgets are being used specifically to gauge the presence and level of pain then it is important to have a good idea of what your cat would normally get up to in a twenty-four-hour period when he or she was fit and well. These are the sorts of things that you should be monitoring.

Sleep, including
- deep sleep
- sleep from which the cat can be easily aroused
- resting with eyes closed

Environmental interaction, including
- looking out of windows
- resting with eyes open
- use of tall observation points

Outdoors, including
- exploring
- hunting
- marking
- foraging
- social interaction or rest beyond the area that you can observe

Social interaction, including
- attention from humans
- grooming or rubbing other cats
- solitary predatory play
- social play fighting
- interactive play with objects
- attention from humans initiated by the cat

Grooming, including
- activity indoors
- activity outside within sight of the owner

Eating

During illness, the above elements of your cat's lifestyle are then monitored in conjunction with other general behavioural observations that potentially relate to the presence of pain.

Changes in character
- fear
- aggression
- reluctance to be handled

Facial expression
- furrowed brow
- hanging head
- glazed expression

Posture
- sternal recumbency (resting upright on the chest)
- 'tucked up' abdomen
- slightly arched spine

Gait
- lameness
- stiffness

Unusual behaviour
- rubbing
- head-pressing

Onset of 'problem' behaviour
- inappropriate elimination (soiling in the house)
- aggression
- localized over-grooming

Vocalization
- changes in sound
- increased/decreased incidence

Sadly I had the opportunity to put this method of pain assessment to the test. During the last forty days of her life, Puddy was carefully and discreetly watched and her deteriorating health became increasingly apparent. She had always been a loving and affectionate little cat, approaching me several times a day with an insistent 'a' for the particular type of slap-loving that she so enjoyed. Within a short period of time this pattern of social interaction stopped and Puddy became silent and withdrawn. Her previous good relationship with the remaining four cats became strained and she hissed and growled at them whenever they approached. I so wanted to hug her and hold her during those last few weeks. Tender loving care is important when your cats are ill, but Puddy was choosing to be solitary and private.

I thought she would have started to sleep more as the condition became more painful but instead she slept less and filled the time by resting, always upright and staring blankly into space. Even in the last few days she continued to eat but only when hand fed, almost as if she was doing it for me and not because she had a strong will to live. Puddy gradually stopped grooming, left off all social interaction with her cat companions and with us and ceased to go outside, but she continued to display brief moments of apparent pleasure in familiar activities. However, by the fortieth day things had deteriorated. Puddy had finally fallen silent and was, by then, in sufficient pain to severely compromise her quality of life. She was quietly and peacefully put to sleep by her vet just before her thirteenth birthday.

It wasn't the easiest thing in the world to monitor Puddy over the last few weeks of her life and sometimes it felt remote and clinical, but I am glad in retrospect that I was able to do so. I have already taken comfort from the fact that I played a positive and compassionate part in ensuring that the timing was right for her. It is never easy to say the final goodbye and not everyone will be comfortable with taking such an active role at the end. Everyone desperately wants their pets to die peacefully in their sleep, but this is a rare situation. Death hardly ever comes peacefully or in moments of sleep. It tends to be messy, degrading, uncomfortable and painful. We shouldn't allow our darlings to end their wonderful lives like that. 'Quality of life' is a much used phrase and it's useful to take time out to really decide what we mean when we say it. I always think of quality of life as being a balance. There is no such thing as a perfect life. If there were, we would all be free from pain, stress and discomfort. Our lives would be one long pleasurefest and we would never know fear, anxiety, frustration or suffering. We would also probably go mad very quickly. How can you judge such pleasure if you have nothing negative to measure it against? So quality of life is a balance between good things and bad things, pleasant and unpleasant. Providing the balance swings in favour of the pleasant things it is possible to say that there is a quality of life. If the only 'pleasant' thing becomes life itself, then there is no longer any justification for carrying on. I will always have this in the back of my mind when the time comes for me to make the decision about Bakewell, Annie, Lucy and Bink.

There are many rituals associated with death that provide a very personal sense of comfort. Some people wish to take a clipping of fur; others need to spend time with the little bodies of their loved ones. It is often helpful to plant a special tree in memory of your cat in the place where he or she most loved to rest in the garden. Some people like to make scrapbooks of photographs to help them to remember the fun and pleasure that the cat brought to the family over the years. Personally I like to make collages of photographs and place them in a clip frame. They hang in my office so my departed cats are never really gone. I also think a beautiful oil painting taken from a photograph can be a wonderful reminder. I have a lovely portrait of Spooky in my bedroom painted by an artist in St Ives. I liked the artist enormously because she asked me about

Spooky's personality so that she could incorporate something of her character into the painting. She certainly did that; Spooky looks so beautiful, and every day I get pleasure from that portrait.

One special ritual I remember was very important to a Romany gypsy I met whilst I was working in a veterinary practice in Cornwall. She had brought a very sick pet rabbit in for treatment and it was soon apparent that the poor mite probably would not make it. As I took him away from her for treatment she asked me to promise I would do something for her if he should die. She believed that the spirits of the dead should be released otherwise they would suffer for eternity. She asked if I would offer his body up to the sky outdoors so that he could fly. Sadly he did die as we expected and, as instructed, I took the body into our small yard and held him up. She was very grateful that I had done it and I must admit there was a very reassuring sense of finality and peace when I did it. When any of my animals have died since I have always given a thought to that gypsy and her rabbit. The skies are very clear in Cornwall and always full of thousands of stars. It is lovely to look up and feel their spirits are all up there. Does that sound daft?

It is always hard to think clearly about things that need to be done at the end. The veterinary surgeon will explain your options (burial, cremation) but it is so difficult to concentrate when you are dealing with the harsh reality that your pet is about to die. I have always planned ahead regarding this issue and it has helped enormously. My home in Cornwall is sitting very securely on a lump of granite so burial just isn't an option for me. I have always elected for an individual cremation (not in combination with other people's pets) and the ashes are returned home in an oak cask. I have even made enquiries at the crematorium regarding my dear old horse Naiad (currently twenty-eight years old and going strong) just to make sure I'm not left with a horrible decision when he dies. Whilst a more ostrich-like approach may appeal to many owners this 'planning ahead' strategy has worked for me. Let's hope I don't have to put any more of these plans into action for a while.

Bereavement and grief

We not only have to deal with the practical implications of our pet's death but also the numbingly painful grief that inevitably follows. Everyone who loses a cat experiences some degree of grief at their death. How we express that grief is a product of our personality, upbringing, life experiences and numerous other considerations, not least of which is the intensity of the relationship we had with the cat who has passed away. For many, a jolly good cry is sufficient to express the pain of the loss of a good companion who gave lots of pleasure. A trip to the local rescue centre some time later will then enable another little cat to share the home of a cat lover. The ease of transition from grief to recovery does not make these people bad owners. They just deal with things their own way.

I don't think I dealt with Puddy's demise in quite such a measured way. After losing Zulu and Bln so quickly, one after the other, and finding out about Lucy's illness I had very few emotional reserves left to deal with the death of my favourite companion. I had almost achieved a pinnacle of grief for my pets beyond which I was unable to go. I had an overriding sense that something good had to come out of such a rotten few months in the history of my cats. I dealt with my grief by writing an article about euthanasia and coping with terminally ill cats. I wanted primarily to tell people about Puddy and to help other owners in similar situations. It was a good feeling when the article was published and people started to tell me that it had helped them get through a very difficult time. The whole experience was so cathartic that I continued to write and that's how this book started. It was my antidote for my own personal grief. I only have the very fondest and most positive memories of Puddy now and I always smile when I think of her.

As grief is a very personal thing it is extremely hard to know whether what we are experiencing when we lose our cats is normal. It is tempting to hide away from things that are painful or worrying, but sometimes it is better to face them. When I was younger I had a recurring dream that involved being chased by a hideous monster. Some well-meaning person told me, when I next had that dream, to turn round and face it in brazen defiance. I did just that, the monster vanished in a puff of smoke and I never had the dream again. Let's face this monster together and see what happens.

Grief

There are several recognized stages of grief that people experience at the loss of a loved one. Human, dog, cat – the process is the same but here I will focus on cats. There is an initial shock and a sense of denial when we first realize that our cat is dead. I remember a distinct inability to concentrate my mind on the concept of death and the fact that I would never see my lovely cat again. The sense of denial was so strong when Hoppy died so suddenly that I kept his little body curled up in a basket for a whole day and just stared at him from time to time before I snapped myself out of it and realized I had to

make 'the necessary arrangements'. After this sense of numb-
ness and denial comes an angry phase. This is a period when
you feel tremendous guilt and wonder whether your cat's death
happened through some omission or negligence on your part.
'Could I have done more?' or 'Should I have spotted the signs
sooner?' or even 'Why did I let him stay out that fateful night?'
Once you have beaten yourself up sufficiently there is then an
overwhelming sense of depression. Everything seems shrouded
in a black cloud and it is impossible to imagine that you will
ever feel happy again. (I've been there!) However, if you are
lucky, and supported by a network of sympathetic friends and
family, you will eventually wake one morning with a genuine
desire to get on with the rest of your life for the first time since
your cat died. This is when you know that, no matter how
much you loved your cat, you have finally accepted your loss.

Many bereaved owners are not fortunate enough to have
supportive people around them and find themselves in a
position of having to cope with their grief in complete isolation.
This is an incredibly difficult task for anyone, no matter how
strong, and I would always recommend reaching out for a
helping hand at this most difficult time. Modern veterinary
practices should be able to provide you with details of local
bereavement support groups; all you have to do is ask. Your
doctor's surgery will be able to provide similar information if
you are finding it difficult to cope. Whoever you speak to will
have been through a similar loss and have an immediate
empathy with you. Never be afraid to ask for help.

Sometimes it's enough to know that the strange physical and
emotional symptoms you are experiencing are normal and that
millions of other people have been there before and pulled
through. It can be frightening when you honestly believe you
are suffering a unique experience.

The expression of grief

The manifestation of grief is a very personal thing and any number of the following symptoms can be experienced as part of the perfectly normal but painful journey to recovery.

Physical signs

- Shock
- Crying/permanent lump in throat
- Shortness of breath/tightness in chest
- Nausea
- Loss of appetite/increased appetite
- Exhaustion
- Dizziness
- Aches and pains
- Disturbed sleep/inability to sleep

Emotional signs

- Sadness
- Anger
- Depression
- Guilt
- Anxiety
- Relief
- Loneliness
- Irritability
- Helplessness

Intellectual signs

- Confusion
- Lack of concentration
- Hallucinations

- Need to talk about the loss
- Need to rationalize the loss
- Preoccupation with death and the after-life

Social signs
- Withdrawal from contact with others
- Rejection of help from others

Just reading through this list I can honestly say that I experienced many of these symptoms after I lost Spooky, Hoppy, Zulu, Bln and Puddy. I distinctly remember hearing Hoppy calling out at night long after he had died. I also saw Bln out of the corner of my eye many times in the first few weeks after his death. I was fortunate enough to understand that what I was experiencing was part of the grieving process and perfectly normal. I also felt terribly guilty when Spooky died because I ultimately had an enormous sense of relief at her passing. She had suffered so badly at the end and I wish I had been braver about calling it a day sooner than I actually did. I also remember feeling angry when Zulu died, focusing on the veterinary surgeon who had attended to him at the emergency surgery. Surely she could have done more? As far as Puddy's death was concerned I think I must have experienced just about every symptom going! However, I'm still here and for those suffering at the moment it's important to remember that there is life after grief.

There are many complicating factors that can exaggerate or prolong grief. For example,

- No previous experience of loss
- Simultaneous loss of a family member
- Not being present at the death

- Witnessing a traumatic or painful death
- Sudden or unexplained death
- Poor coping skills generally
- Lack of support or insensitive comments from others
- Comments from others that trivialize the loss

These are just a few examples but all can hinder the process. There is no structure or set pattern to the journey through grief. Some owners move backwards and forwards through various stages or get stuck in the angry stage for an impossible length of time. Whatever happens, the advice is to accept that grief is inevitable and to succumb to it. Problems can occur when you don't allow these feelings to express themselves fully. Unresolved grief is a common problem in Western society where it is traditional in many countries not to express intense emotion publicly. Many times I have been with owners as their pets have been put to sleep and their tears afterwards have been for many other reasons, not just their lost cat. It often appears to me that we are a society more tolerant of public displays of grief at the passing of a pet than at the death of people.

Eventually the sadness and the tears go. They are replaced by chuckles and smiles as we remember all the happy times we had together. That's why we have cats. It's not about the pain of losing them, it's about the pleasure we get from each other when they are alive!

Coping with bereavement
- Plan ahead and have an idea in your mind what arrangements you wish to make at the end: burial, cremation, return of ashes, etc.
- Accept that grief is perfectly normal.
- Accept help from family and friends and don't take to heart

anything that is said by someone who doesn't understand owner/pet relationships.

- If you are relatively isolated in your social life don't be afraid to talk to staff at your veterinary practice, your vicar/priest or a professional bereavement counsellor about your feelings.
- Make a scrapbook or have your favourite photograph of your pet enlarged and framed; it may help.
- Try to ensure you are looking after yourself whilst you are grieving; it won't help the pain to go away if you stop eating.
- Allow yourself time to mourn but try to return to normal routines as soon as possible.
- Understand that recovering from grief does not mean you forget about your lost cat or love them any less.
- Considering the acquisition of another cat is not disrespectful to the memory of the departed. It's a compliment!

❉   ❉   ❉

Life continued after Puddy's death but I admit spending some time in a state of deep depression. I am fortunate enough to have a job that enables me to make a difference to other people's lives and this was extraordinarily therapeutic. I continued to help others to improve their cats' well-being and it wasn't long before my naturally positive and upbeat attitude returned. However, not a day goes by when I don't think of Puddy and all the pleasure she brought into my life. I miss her very much.

Epilogue

SO, THE STORY SO FAR IS OVER. YOU HAVE MET THE NINE CATS I have shared my life with and I hope their stories have helped you in some way to better understand your own cats. Bakewell, Annie, Lucy and Bink are still very much with us and going strong. Bakewell has now recovered from his recent illness and Lucy is stable on her medication for FIV. Annie is still very stiff in the mornings with her arthritis and we are trying to find the best treatment for her at the moment to control the pain of her condition. She always manages a hearty head-butt when she sees me so I figure she isn't feeling too bad. Bink is still weird but I swear she is mellowing with age!

Despite all my efforts over the years I cannot say I will ever fully understand the cat. The more I know the more questions are raised, but I continue to seek experience and knowledge to allow me to go on helping cats have as pleasant a life with humans as possible (and vice versa, of course). The really challenging thing about working with cats is their complete lack of uniformity. None of them read the textbooks that say how they should behave in certain circumstances. My aunt's cat Bo-Bo lived to be twenty-one years old on a diet of raw liver, single cream and Caramac chocolate. Nobody ever told him that he should have popped his clogs about eighteen years previously on such a regime. Seriously, don't try this one at home! Bo-Bo survived on heavy supplementation with mice.

Every time I give a presentation or talk to cat owners I am careful about making statements as cast-in-stone facts. If you ever say 'cats don't like tin foil' there will always be someone who says their cat likes nothing more than to sleep curled up on a sheet of Bacofoil. When I remark that cats don't like to be sprayed with jets of water I am bombarded with contrary opinions about cats loving showers, baths and swimming. I don't state facts any more; I generalize. It's much safer. There is no such thing as a single standard programme for urine spraying or a quick fix for inter-cat aggression. Every situation is unique because every cat is unique. So if there are any generalizations in this book that you disagree with, you now see why. Cats are not machines that can be mended by looking at the operator's manual; if that were the case they wouldn't be half as much fun.

In my ongoing quest for knowledge and enlightenment, I do find something a little disheartening. As I start to explain the reasons for certain behaviour in my patients I often see a look of utter dismay on the owners' faces. My exploration into the reality of the nature of the cat often dispels the popular myths that have formed the basis of their relationships. Many of the stories that people tell me are considered proof of their pet's extraordinary sixth sense. What they are mainly describing is the heightened levels of the other five. For some people a better understanding actually destroys an essential element of the perceived bond and I feel a real sense of denial and refusal to accept the truth. There is a worry that the more we know the less enigmatic and magical the cat will become. Never! I can honestly say that my knowledge (or understanding of what I believe to be true) has actually enhanced my relationship with my cats and made the whole inter-species thing even more wonderful. Every day that goes by I learn another amazing fact about the domestic cat. I also discover how little we all know about the animals that share our lives. For many the relationship is mere serendipity as the cat says one thing, the human says another, nobody understands but somehow it seems to work. I would like to think that at least I am trying to learn the language.

The profession of pet behaviour counsellor is very trendy at the moment. Over the past few years many newcomers have entered the realm via accredited courses and certificates. How effective they all are will vary greatly and depend on their theoretical knowledge and their practical experience, but most of all it will be governed by their compassion for the owner. Let's face it, the popular conception of the 'cat on the couch' just doesn't happen. The effectiveness of behaviour therapy is as much about the relationship between the counsellor and the owner as anything. This, above all else, is the thing I cherish most about my work over the past few years. I have a deep and genuine regard for all my clients. They have been, almost without exception, kind and honest people. Many have also had the most extraordinary sense of humour and I hope they enjoy the way I have chosen to write this book. I have been privy to a lot of secrets over the years and I think it's important to reassure everyone that those secrets remain just that. All the stories I have told within this book are based on real cases. I have, however, disguised them carefully by mixing and matching elements of several, to ensure that nobody would feel betrayed in any way. My consultations will always remain confidential. The only reason why expurgated versions have been discussed is purely to help others.

I would love to feel that I could conclude my story by offering the ultimate pearl of wisdom about cats. Maybe a clever little phrase or statement that would make every cat lover nod wisely and say, 'Oh, yes, I know what she's saying.' I've thought about this long and hard and I have concluded that it would only end up being another one of my generalizations. However, there is one thing that has cropped up time and time again in my work over the years. It's probably an element that everyone would be surprised I even get involved in but it's definitely there, and those owners that suffer from this problem accept it can be absolutely devastating. The interesting thing is that it will probably be the only remark that I address specifically to men. As you are probably aware by now, most of my clients are women. Most cat owners are women. That doesn't mean that every good cat owner is female – look at Peter. However, just to provide a little more balance to the proceedings, here is a piece of advice for men only.

Never, ever, say, 'It's the cat or me!'

This is the one battle of the sexes that the man is sure to lose every time. I have often been asked to resolve relationship issues between man and cat. Mainly, either *My cat hates my new boyfriend/husband/partner* or *My cat is frightened of my new boyfriend/husband/partner.*

I remember one case where the boyfriend (soon to be husband) was so desperate to appear committed to the care of his girlfriend's beloved cat that he dutifully agreed to hold him whilst the cat received a worming tablet. Unfortunately his overzealous desire to please resulted in a vice-like grip that terrified the cat so completely that it suffered from major post-traumatic stress disorder every subsequent time he saw the man. I was duly called upon to put things right and it was clear that the boyfriend was keen to make amends. I feel a little guilty now but I asked him to do a number of things that a strapping rugby player would normally find quite repulsive. He had to walk quietly, talk in a high-pitched voice, feminize his body language, wear his girlfriend's clothing and generally sacrifice himself on the altar of embarrassment. Such was the devotion of this particular guy that he was prepared almost to forfeit his masculinity to achieve the love of that cat. It worked and the couple were soon enjoying an innocent *ménage à trois* as the cat changed his mind. He had obviously decided that his new daddy was brilliant and that there was nothing better in the world than a good cuddle at night. The only unfortunate thing was that the best man chose to relate the whole tale, complete with gory details, as part of his speech at the wedding.

Occasionally I get involved in things that I probably shouldn't. One call I received was from a very unhappy lady who reported that her elderly cat was not settling in well to the new marital home after her recent wedding. Basically, her new husband hated her cat and she was powerless to change his mind. The two 'people' she loved most in the world were at war. I honestly don't know what I expected to be able to achieve but the lady was so insistent and so depressed I couldn't really say no when she asked me to visit.

She was waiting for me at the door when I arrived and

seemed very relieved to see me, if somewhat nervous. She ushered me into her sitting room to meet her husband. At first sight he appeared to be a broadsheet newspaper with a pair of crossed legs attached. He would not lower his *Daily Telegraph* and he would not acknowledge his wife's introduction. I must admit I sat down wondering why I had bothered sitting in traffic for two hours to get that response but I was definitely curious. A man prepared to be that discourteous probably had a very big issue about something. His wife left the room to make a cup of coffee and I turned to him and said, 'Daft, isn't it, being a cat shrink for a living?' He lowered the paper only long enough to tell me that he was going out to play golf. Well, by fair means or foul I managed to delay his game. We talked about golf, cars and motor racing. I was determined not to talk about cats at all until I was sure he was on my side. His wife had reappeared briefly and, seeing that things appeared to be improving, left the room again to try to find Chiquita, her little cat.

Gradually the husband's body language relaxed sufficiently for me to feel I was in a position to change the subject to the dreaded taboo. After all, how difficult could it be? It was only one small letter change – from R to T. He might not even notice. So my next sentence went from car to cat in a fairly smooth transition. As I suspected, his hatred for Chiquita wasn't really hatred at all. He was absolutely terrified of her and completely unable to admit this to his beautiful new wife. He was a big strong man; how could he possibly admit that he was frightened of a small furry creature? He told me that when he had first met Chiquita he had tried to ignore her, hoping she would go away. However, this was guaranteed to display the most attractive body language to Chiquita and she immediately jumped on his lap and put her little tabby face right into his.

His description of the horror he felt was incredibly vivid. Here was a true ailurophobe. It soon became clear that this poor man had experienced a rather unpleasant childhood encounter with a cat when he was seven years old. He had seen a little ginger cat sitting on a wall and approached for a stroke and a friendly cuddle. What he got was a bunch of claws across the face for his trouble. You and I know that this was merely the act of an individual cat reluctant to indulge in interaction with a human. He may have been poorly socialized with people or he may have been fearful. Unfortunately, this single event had shaped one person's perception of cats for ever. They were vicious, unpredictable and not to be trusted.

I felt a great deal of sympathy for this man. Whatever we all feel about cats there are many people with a genuine fear of them. I am not a professional human counsellor or psychologist and I have no training in dealing with human phobias. I explained this to him and his wife (I was delighted that he had eventually decided to tell her about his phobia). After some discussion it was apparent that it really didn't matter that I couldn't help. Whilst I was there I justified my fee by dis-cussing Chiquita's diet and lifestyle and various suggestions for making her life more fulfilling and enjoyable. I also spent a little time explaining to the husband why that ginger cat had been so mean to him all those years ago. Everyone seemed relaxed and happy when I left; I had a strong sense that this would not be a big problem for much longer. Over the next few weeks we talked on the telephone about a gradual improve-ment. He still wouldn't touch Chiquita but he was able to stay in the same room without feeling anxious. By the end of the eight weeks he was acknowledging her when she came in and even felt that it wouldn't be long before he could stroke her. He was actually a very nice man; it's quite worrying how you can make an assumption about a person on first impression that is nowhere near accurate. I often wonder how different things would have been if Jenny had been like that little ginger cat. I probably would be an accountant.

Jenny definitely decided that was not to be and I became a cat behaviour counsellor instead. Over the years I have sat in many traffic jams, I have drunk many cups of tea and I have crawled across many carpets in the pursuit of urine. And now, despite the fact that I love my job, I have got to the stage where I have become frustrated by the inescapable truth that my life revolves around picking up the pieces when things go wrong. I can never see a time when I stop consulting for good (I think after all this time it has become an intrinsic part of me) but I like the idea of preventative behaviour therapy rather than curative. It's a very exciting concept that problems can be avoided by knowledge and understanding and it has become a challenging project to spread the word. It is still hard to believe that I have embarked on that journey without real intention. I started writing this book mainly as a personal therapy to help me cope with the loss of so many of my cats in such a short space of time. It was a private musing initially and a great way of exploring my own relationship with my wonderful cats. I can't quite remember when it happened but, at some point, I suddenly realized that I could have something useful to say to everyone else who loves cats. Why not share it? People I meet in the course of my work all remark that I really should write a book and it seemed that the time had come to do just that. No more excuses! So as I continued to write, it became not just a story but also a preventative manual that contained possible answers for many (but definitely not all) questions that cat owners may ask. However, more important than that, it had metamorphosed into an exploration of the fascinating emotional relationship that can occur between (wo)man and cat. This to me is the truly exciting part of my brilliant job.

It is very hard to find the right way to finish a book that has become, in itself, such an important turning point in my life. It's difficult to find the words. After some reflection I have decided to end with a short tale that perfectly illustrates every-thing I have been trying to say.

My dear friends (you know who you are!) have always loved cats. Several years ago their last remaining cat died and they decided that they wouldn't have another. They were shortly to retire and felt that the obvious restrictions that came with the responsibility of pet ownership would hinder their plans to travel and see the world. For one reason and another we lost touch for a few years until one day I found myself con-sulting at a house only a street away from their home. I decided to pop in and see them (I actually wasn't even sure they still lived there) and luckily they were in and delighted to see this face from the past. We chatted briefly and arranged to see each other the following Sunday.

When I arrived I was amazed to see a frail and tatty old cat curled up tightly on a thick crimson cushion positioned on the sofa. She was fast asleep and apparently deaf as a post and oblivious of everything around her. I was quick to remind them of their promise that they would no longer have cats because of their travel plans. They were even quicker to point out that she wasn't their cat. Apparently she lived in the house opposite and had recently wandered over the road and expressed a desire to enter their house. They duly obliged and she seemed par-ticularly keen on the contents of their fridge. That's how they discovered her taste for prawns. Once she had eaten sufficient she made it obvious that she wished to position herself on the sofa. They lifted her up and placed her gently down on the seat.

She soon made it clear that she required something a little more padded, hence the crimson cushion, and she curled up to sleep for the rest of the afternoon. From that day forward she arrived at eleven o'clock every morning. There she remained until six o'clock in the evening when she would look up and give a plaintive howl. My friends would then, with some ceremony, lift the crimson cushion and walk slowly, side by side and in a sombre fashion, across the road to their neighbour's house with their precious cargo reclining majestically on her throne. She would then receive her evening meal and settle down for the night until eleven o'clock the following morning when she would call at my friends' house for prawns and to repeat the process all over again. Needless to say my friends don't do much travelling now. How can they if they always have to be home between eleven in the morning and six in the evening?

Quod erat demonstrandum!

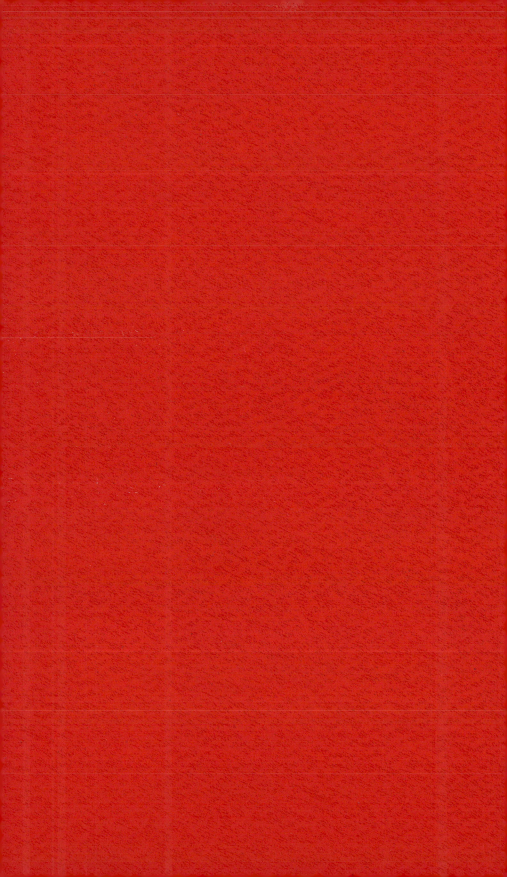